The People's Forests

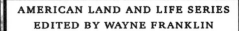

AMERICAN LAND AND LIFE SERIES
EDITED BY WAYNE FRANKLIN

The People's Forests

ROBERT MARSHALL

Foreword

Mike Dombeck

Biographical Essay

Douglas K. Midgett

University of Iowa Press ⴘ Iowa City

The veneer referred to on page 31 of this book refers to the hardcover,
1933 edition rather than to the current paperback edition. Chapter 1 of
the original edition began on page 13. No text from the original book
has been omitted from this edition.

The publication of this book was generously supported by the
University of Iowa Foundation.

Printed on acid-free paper

Library of Congress Cataloging-in-Publication Data
Marshall, Robert, 1901–1939.
The people's forests / by Robert Marshall; foreword by Mike Dombeck;
biographical essay by Douglas K. Midgett.
    p.   cm.—(American land and life series)
Originally published: New York: H. Smith and R. Haas, 1933.
Includes bibliographical references (p.).
ISBN 0-87745-805-7 (paper)
1. Forests and forestry—United States.   2. Forests and forestry—
Social aspects—United States.   3. Land tenure—United States.
4. Forest conservation—United States.   I. Midgett, Douglas K.
II. Title.   III. Series.

SD143 .M35 2002
333.75′0973—dc21   2002019849

02   03   04   05   06   P   5   4   3   2   1

Contents

The People's Forests in the Twenty-First Century
BY MIKE DOMBECK VII

Bob Marshall: The Call to Activism
BY DOUGLAS K. MIDGETT XV

Acknowledgment

I. Devastation Sweeping the Country 13

II. Forest Values—Raw Material 23

III. Forest Values: Soil and Water Conservation 43

IV. Forest Values: Recreation 57

V. Forests and Human Happiness 77

VI. Private Ownership 87

VII. Public Regulation 109

VIII. Public Ownership 123

IX. Method of Acquisition 141

X. Land, Labor, and Livelihood 159

XI. Recreational Safeguards 171

XII. A Biological Interlude 189

XIII. A National Program 209

References 221

Index 225

The People's Forests in the Twenty-First Century

MIKE DOMBECK

As we embark on the twenty-first century and human populations continue to increase, we face daunting conservation challenges. One hundred years ago Secretary of Agriculture James Wilson directed that the forest reserves, now called national forests, be managed "for the greatest good for the greatest number in the long run." This famous quote was often invoked by Bob Marshall as he fought for wilderness preservation some seventy-five years ago. Much has changed since the days of Wilson and Marshall, but many challenges remain. Our technological and industrial society of 275 million people is placing unprecedented demands on the land, all the while remaining dependent upon it. Just as our earliest ancestors depended on the land for survival, so do we depend on it for our well-being and our quality of life. Managing land with an eye toward long-term benefits has proven to be difficult. Aldo Leopold, famous author of *A Sand County Almanac* and a Marshall contemporary, wrote that the "oldest task in human history" is "to live on a piece of land without spoiling it." The following are some of the key public land challenges we face at the dawn of this new century.

Water

The *International Herald Tribune* recently reported that two-thirds of the world's human population will face water shortages by 2025. About one-third of the acreage in the United States that is covered by forests produces about two-thirds of the runoff. The cleanest water in the country flows from our forest lands; the very cleanest of that water flows from the wilderness forests, roadless areas, and undeveloped landscapes that Bob Marshall dedicated his career to preserving. Collectively, our public lands are by far the largest and perhaps most important water provider in the United States. The 192 million acres of national forests provide drinking water to more than sixty million Americans living in some 3,400 communities in thirty-three states, at a marginal value of more than $3.7 billion per year. Ironically, forest policy discussions commonly include the value of wood, minerals, oil, and gas but rarely the value of water.

A watershed is all the land drained by a single river system; watershed function is the interaction of soil, water, and vegetation in that system. Small watersheds drained by a single small stream are nested within a larger watershed. Given the fundamental importance of water to all life, watershed health and water quality should be the basic measures of success for our public land managers.

Sprawl and Development

Development and urban sprawl are occurring at an unprecedented rate. Between 1992 and 1997, an average of 8,700 acres per day of wild land, farmland, and open space were turned over to developers. That is more than double the rate of the 1980s. The number of privately owned tracts of forest land containing less than fifty acres doubled from 1978 to 1994. Highly developed and paved-over landscapes maintain poor watershed function and provide only limited basic ecosystem services. The basic challenge is, how will our society slow the sprawl?

Old-Growth Forests

Old-growth forests symbolize, more than any other issue, the conflict and controversy that have crippled the Forest Service for more than thirty years. With only a small percentage of our old-growth forests remaining, it makes little sense to harvest them simply to bring their short-term economic values to market. The greatest good of these remnant forests is found through research and study, conservation and restoration. The maturity of a truly wealthy nation is not measured in acres harvested, rivers dammed, oil barrels filled, or mountaintops mined; it is most ably displayed by demonstrating mastery over ourselves. Our willingness to say, "Enough; these ancient forests cannot be improved through commodity timber production," honors our nation far more than does engineering an expensive road to harvest an old-growth stand. Millions of acres of previously harvested forest, particularly in the fire-dependent ecosystems, are in need of thinning and prescribed fire. These processes provide both wood fiber and high-quality jobs for thousands of people. Timber harvest is important to our society, but must we cut every acre? Ensuring the conservation of the few remnants of old-growth forests should be among the highest priorities for our national forests.

Roadless Area Protection

Like old-growth forests, roadless areas have been a major source of public debate and scrutiny. In the second Clinton administration, efforts to keep millions of acres of national forest roadless met with very strong public support. Unfortunately the Bush administration, bowing to industry and political pressures, has chosen to revisit and in some cases weaken policies governing roadless areas.

Due to complexity, cost, and controversy more project proposals fail in roadless areas than anywhere else in the national forests. Additionally, it is hard to believe that a single private landowner would continue to build new roads in pristine areas if they were faced with the combination of a crumbling 386,000-mile road system and an $8.4 billion road maintenance backlog. Local communities want local decisions. This begs the question of who is willing to accept this multibillion dollar taxpayer liability that is the cumulative effect of hundreds of local decisions. Who is responsible for this nationally subsidized road system?

One-quarter of one percent of our nation's timber and a

fraction of a fraction of our oil and gas is a small price to pay for the protection of fifty-eight million acres of our children's natural resource inheritance. The long-term public interest in conserving these areas should prevail over short-term interests.

Off-Road Vehicles

More recreation takes place on national forests and grasslands than on any other type of public land. Privately owned land is increasingly posted with "no trespassing" signs and is off limits to most people. Technological innovations in motorized recreation enable people to get into more remote areas than ever before, often resulting in degraded water quality, degraded wildlife habitat, and erosion. Off-road vehicles can make legitimate use of public lands where they are expressly allowed, but we must ensure that their use does not compromise basic watershed function and wildlife habitats.

The 1872 Mining Law

Hard rock mining on public lands is governed by a statute that was passed when the country was distracted by Civil War reconstruction, a time when women could not vote and when St. Louis represented the western frontier. The 1872 Mining Law confounds Forest Service and Bureau of Land Management efforts to balance multiple use by allowing privatization of public lands for as little as $2.50 to $5.00 per acre. Every single use of the 464 million acres of Forest Service– and BLM-managed public lands—recreation, timber harvest, oil and gas development, for example — is subject to the approval or rejection of a field official for environmental or safety reasons. Every use but one, that is: hard rock mining. This law vests multinational companies with the right to mine an area simply by virtue of the presence of valuable minerals. Many wild places are threatened, such as the portions of the biologically rich Siskiyou National Forest in Oregon, the Cabinet Mountains Wilderness in Montana and the Greater Yellowstone ecosystem. Congress needs to update this dangerously archaic law.

Loss of Biodiversity

Wilderness, roadless, and wild lands in general serve as biological refuges to many native species. National forests and grasslands have, for example, 181 of the 327 watersheds identified by the Nature Conservancy as critical for the conservation of biodiversity in the United States. National forests and grasslands support over 350 species of plants and animals federally listed as threatened or endangered, plus another 2,800 sensitive species and numerous imperiled plant communities. High biodiversity enhances ecosystem stability, resilience to natural disasters such as fire and flood, and resistance to invasion by non-native species. How do we stem the tragic and increasing loss of species?

Private Land Conservation

Few areas offer more promise for conservation and watershed restoration than private land. Some of the nation's largest wood retailers have recently decided not to purchase wood from endangered forests and to sell only appropriately certified wood products, an action offering hope for the conservation of our forests. The National Research Council, however, has reported that of the 9.9 million private woodland owners in the United States only about 5 percent have professionally developed science-based management plans. Programs that offer voluntary options for improving management, conservation, and restoration of private lands are chronically underfunded.

Wilderness

Few decisions are more forward-looking than those involving wilderness designation. At the time of this writing we have 104 million acres of congressionally designated wilderness, but it remains constantly threatened by mining, invasion of exotic species, and overuse, to mention only a few. Remaining vigilant against these threats is critical. For our country's long-term interests we should be recommending the expansion of wilderness—from remote high-elevation areas to old-growth forests, prairie grasslands, and bottomland hardwoods—to ensure that all major ecosystems are represented in our wilderness preservation system.

The stated conservation issues, and many more not mentioned here, would be more pressing if it were not for our currently preserved wild lands. Bob Marshall clearly understood this even though the passage of the Wilderness Act came decades after his untimely death. The Wilderness Act has a soul, an essence of hope, a simplicity, and a sense of connection. Unlike the jargon-filled tomes of most laws, the Wilderness Act says in very few words that some of what we have today is worth preserving for future generations. Bob Marshall's life was also dedicated to keeping wild places wild.

Note

Portions of this essay were taken from a March 27, 2001, letter written to Secretary of Agriculture Ann Veneman by U.S. Forest Service Chief Mike Dombeck.

Bob Marshall:
The Call to Activism

DOUGLAS K. MIDGETT

When Bob Marshall flew to Wiseman, Alaska, in August 1938 on his third trip to the village, he was dismayed to find the changes that had taken place since his previous visits less than a decade earlier. Planes, tourists, and motor vehicles had arrived to compromise the isolation that had so captured him during his earlier stays. Already the place Marshall first saw as "two hundred miles beyond the edge of the twentieth century"[1] had been penetrated by a modernity from which he sought respite.

We can only speculate what his reaction would be to the present situation. In 1974, work began on a road that would link the Eliot Highway north of Fairbanks to the Beaufort Sea at Prudhoe Bay on Alaska's north coast. Five months later this graveled surface, later named the Dalton Highway, was completed, providing access for the heavy equipment necessary to sustain the 800-mile Trans-Alaskan Pipeline. The road parallels the pipeline for 400 miles, passing a couple of miles from Wiseman and breaching the wilderness that Bob Marshall loved. Travelers on the Dalton Highway near Wiseman gaze west across an expanse that contains the Gates of the Arctic National Park (the name is Marshall's creation), but in the foreground they cannot escape the presence of the four-foot-diameter pipeline as it snakes north to the oil fields. Nor can they avoid other intrusions like the Alaska state troopers who patrol the length of the highway, the helicopters engaged in constant surveillance of the line, and the occasional pumping stations that keep north-slope oil flowing to fulfill the nation's appetite.

Marshall's Alaskan sojourn, undertaken in two trips in 1929 and 1930–1931, was the singular experience that helped him define his ideas about the value of wilderness and the necessity of maintaining areas of unaltered nature. He had been an avid outdoorsman and hiker since boyhood, exploring New York's Adirondack region with family and friends (he was one of the original three people to climb all the highest peaks in the Adirondacks). He had experienced the Northern Rockies as a Forest Service employee stationed in Missoula, Montana, between 1925 and 1928. He had already penned two articles arguing for the unique qualities of wilderness experience and decrying the depredations being exacted on wild areas.[2] But it was in Alaska that he sought and found the kind of isolation and wondrous untouched nature that would animate his thoughts and writings for the rest of his life. During most of the 1930s, following his second trip to Alaska, he would devote his time to an advocacy for wild areas and for responsible forest management that is perhaps unmatched in the American conservation movement.

Although a work long out of print, *The People's Forests* carries a significant message for the present day. In this introduction I focus on two aspects of Bob Marshall's life's work—his conviction that public ownership and management of the nation's forest resources were imperative if they were to be saved as a resource for future generations, and his abiding dedication to the preservation of wild areas. These two thrusts are inseparable in the thinking and writing of Bob Marshall, and it is certain that he believed that the latter was not realizable without a program that would address the former.

In constructing this portrait I locate Marshall's thought and action in the context of his times, of the intellectual and social ferment that contributed to the formation of his extraordinary and prescient vision. In this context I focus on two recurrent themes that animated his writing: his attraction to socialism as a mode of social organization—particularly as it applied to a patrimony that he recognized to be fragile and subject to the depredations of human exploitation—and his understanding of the psychic cost of life in a modern, industrial, urban environment. That these were profoundly part of his consciousness is borne out in much of his writing.

Intellectual and Ideological Roots

Bob Marshall's family, his social position, and the ideas that constituted his early experience cannot be dismissed in a consideration of the man he became. His father, Louis Marshall, was a prominent lawyer of national status who was often in the forefront of issues of civil rights and the struggle against discriminatory abuses. The Marshalls were Jews. Louis Marshall, the son of immigrants, has been described as a "life long Republican,"[3] but in the first decades of the twentieth century the most prominent Republican was Theodore Roosevelt, a progressive who believed in the necessary role of government in curbing the abuses of unfettered capitalism and who, as creator of the Forest Service, is forever associated with the cause of environmental conservation in this country.

Louis Marshall's commitment to the rights of minorities and his concern about the anti-Semitism rampant in the United States of the early 1900s occurred in a time of congressional hearings that preceded the enactment of the nation's first immigration acts, hearings that gave voice to the worst of "scientific" racist claptrap. It was a time of the infamous "Palmer Raids," a period of national chauvinism and jingoism, when to be adjudged "different" from Anglocentric conformity and to engage in political dissent was to invite authoritarian assaults. His earliest educational experience suggests exposure to thought that would contribute to an enlightened social awareness, for he was sent to the Ethical Culture School in Manhattan, Felix Adler's experiment in the education of the poor. The school taught social values and stressed the importance of social

commitment and service while providing a rigorous aca-
demic experience. Given this background, it is not surprising that Bob Marshall evidenced a refined social consciousness at an early age.

Louis Marshall also bequeathed to Bob and his other two
sons an abiding concern for wilderness preservation. Years
before Bob's birth Louis Marshall's efforts to protect the
wilderness qualities in what became the Adirondack Forest Preserve signaled a dedication to the cause that his sons
would follow and that Bob would make his life's work.[4]
Devotion to the preservation of Adirondack wilderness was
the first of Bob Marshall's great efforts in the service of wild
areas.[5]

Bob was to become more than an enthusiast, more than
an advocate. From his early experiences at his family's summer home in the Adirondacks he determined to be a professional forester. He completed his B.A. at the New York
State College of Forestry at Syracuse, continued at Harvard
Forest where he did the work for his M.A., and finally took
a Ph.D. in plant physiology from Johns Hopkins in 1930. In
between the latter two degrees he worked for the U.S. Forest Service for three years at the Rocky Mountain Forest
Experiment Station in the Forest Service's Northern Region
where he had the opportunity to stretch his legs in mara-
thon hikes through some of the most pristine alpine wilderness remaining in the forty-eight states. It was also an
experience that sharpened his awareness of social inequal-
ity in American life and his commitment to a socialist per-
spective.[6]

When, after completing his Ph.D., Marshall embarked

on his second trip to Alaska, it was with the intention to remain for a year to explore the vast uncharted area in the Brooks Range and the Koyukuk River drainage. As it turned out, he was to do an insightful ethnographic study of society in this remote frontier setting.[7] That his Alaskan experiences contributed in large measure to his advocacy for protected wilderness is unquestioned: He would gladly have placed most of the territory locked away from the exploiters and despoilers. This, of course, did not happen, but Marshall's Alaskan sojourn sharpened his perceptions, intensifying his commitment to the cause that had already assumed the central focus in his young life.

After his year in Alaska, Bob Marshall was enlisted to write sections on forest recreation for the monumental Copeland Report—the 1655-page comprehensive study of the nation's forests undertaken for the U.S. Senate—known officially as *A National Plan for American Forestry*.[8] His sections of the report expressed in detail his understanding about the need to protect a variety of lands for the purpose of future public enjoyment. They also expanded parts of a publication that he had written in 1930 for the League for Industrial Democracy entitled *The Social Management of American Forests*, containing his first comprehensive examination of the need for public management of the nation's forest resources.[9]

In the section of the Copeland Report entitled "The Forest for Recreation," Marshall commenced with a discussion of forest recreation usage citing voluminous figures and facts on public usage, present and future. It was typical of his approach; he was an inveterate gatherer of facts and

statistics, and he was devoted to the factual demonstration of a position. He cited the numbers of visitors to national parks and national forests over a fifteen-year period,
demonstrating the extraordinary growth in recreational use
during that time. He calculated future increases based on
population growth, reduced work time, rising standard of
living, transportation improvements, and "increasing psychological necessity for escape to the primitive."[10] Concerning the latter point he elaborated: "As society becomes more
and more mechanized it will be increasingly difficult for
many people to stand the nervous strain, the high pressure, and the drabness of their lives. To escape these abominations, increasing numbers will seek the primitive for the
finest features of life."[11]

This passage echoes some of his earlier statements on
the psychic value of visits to wild areas. "Civilization," he
said in his 1930 *Scientific Monthly* article, "requires most
lives to be passed amid inordinate dissonance, pressure and
intrusion. . . . [T]he chance of retiring now and then to the
quietude and privacy of sylvan haunts becomes for some a
psychic necessity. . . . [It] saves them from being destroyed
by the terrible neural tension of modern existence."[12] In
Social Management of American Forests, he wrote: "Under
our present civilization, with its great amount of dreary,
monotonous labor, with its incessant invasion of privacy,
with its grime and sordidness, with its almost unbearable
cacophony of nerve-wracking noises, it may become a matter more important than any practical consideration to
enable people to enjoy the adventure, the freedom, the
cleanliness and the repose of the forest."[13] A passage in the

Copeland Report celebrates the exhilaration experienced when one deals with nature unmediated by artifice: "When a man travels in the forest by natural means all his capability and all his understanding are called into use, and an infinite number of subtleties . . . are clearly appreciated and become vitally significant to him."[14]

These expressions are replicated numerous times in Marshall's writing. Written at a time when American intellectuals were animated by the writings of Freud and exploring the possibilities psychoanalytic theory held for understanding all manner of human behavior and organization, these passages reveal influences common to the intellectual milieu in which Marshall was immersed. Echoes of *Civilization and Its Discontents* and Marx's concept of the alienation of labor emerge throughout Marshall's published writings and correspondence. The interest in psychoanalytic tools prevalent in American anthropology in the 1920s finds expression in Marshall's Alaskan research, wherein he administered IQ tests and other projective instruments to the residents of the Koyukuk region.

Alaska and the Maturation of a Wilderness Ethic

Bob Marshall encountered in the Koyukuk country people whose vision of the world was in substantial accord with his own. He attended to their conversations, particularly those who spoke of corporate greed, the fetters of capitalism and the strangulation of individual freedom through the wage form. "I see they figure there're ten million people unemployed now—and thirteen percent of the people own ninety per cent of all the wealth of the country," he recorded in *Arctic Village*. "I don't believe in socialism myself . . . but I'll be God damned if this private greed isn't ruining the country," he wrote down a few pages later. And in summary: "'Outside': they are only half-awake—dull, routine slaves, tied down to follow each other."[15]

Marshall also recognized in Wiseman a form of organization one might term a "communal" mode of production, where little social distinction is made between employer and worker, because the relationship may be reversed months later when another task is undertaken. This underpins an economic system characterized by "the old-fashioned rugged individualism of the frontier." Marshall contrasted this with the "rugged individualism" which received so much attention in the Republican propaganda of 1928. Concerning that type, one of these true individualists of the Arctic remarked: "This rugged individualism they're talking about Outside, it's ninety-nine individuals do the work and one individual gets all the profits."[16]

In keeping with the communal aspects of Koyukuk society he described the importance of voluntary labor and the significance attached to the work of each individual. This communal focus appears where a profit motive is secondary to the importance of maintaining the mutual well-being of the village. He observed that to charge what the market might bear in a time of scarcity "would be taking advantage of a communal misfortune for one's own aggrandizement, and this would be considered the most indefensible sort of avarice."[17] This attitude has its foundation in a social order where all are accorded worth and respect in keeping with their contributions: "Every person feels that he has an important role in the communal life, and every person feels that all the other people are likewise significant."[18]

Marshall also recognized dangers to this Edenic existence. Although the townspeople envisioned further development of the area's resources through outside investment, Marshall sounded a warning note: "If Outside money were to come into Koyukuk in any large amount and absentee ownership be established as the rule instead of the exception, most of the unique freedom and independence which distinguishes the individualistic economic system of the Koyukuk would disappear."[19]

Finally, Bob Marshall gave voice to the residents on the subject closest to his own heart—their love for the natural world that surrounds and sustains them.[20] It is here in *Arctic Village* that we find clues to the emerging vision that he would carry throughout his efforts during the next decade, for he recognized that even in this remote place the possibility of environmental despoliation was not only possible but also palpable. Moreover, in this experience he developed a sense of the scope of the solution necessary in order to solve this problem. As Roderick Nash notes: "He was the first American to recognize that wilderness preservation in Alaska could involve whole watersheds, entire mountain ranges, and intact ecosystems."[21]

The idea of wilderness preservation Marshall first clearly articulated in "The Problem of the Wilderness" had become his own paramount focus, but he realized it would be best addressed in social programming as the nation came to grips with the human devastation caused by the Depression. Upon returning from Alaska he embarked on his contribution to this task.

The Socialization of American Forests

Given his socialist orientation and his passion for forestry as a profession, it is hardly surprising that Bob Marshall would join these two intellectual strands in his concept of forest management. In 1930 he had proposed a plan for public control and administration in *The Social Management of American Forests*, in which he made a strong case, once again bolstered by an assembling of statistical data, for socialization of the nation's forest resources. The argument rests on the importance of forests for the realization of various values and on Marshall's demonstration of the devastation caused by private ownership and mismanagement. Despite his belief in the advisability of social control, he was under no illusions about the possibilities of effecting these changes in the short term, observing that "it is extremely unlikely that socialization of forest land will be undertaken on a large scale for many years to come."[22] Nonetheless, he recognized areas where governmental policy could be maintained and extended, particularly in recreation and streamflow regulation. He was, however, unwilling to concede to private owners the continued right to ruin forest resources in the sole interest of profit-taking:

> Such inability of the private owner to meet the requirements of both public welfare and private profit should certainly not be relieved by leniency in the enforcement of

regulations necessary for the protection of public needs or by subsidy made at public expense. Rather the private owner should be forced to sell out to the governmental agencies which for a quarter of a century have been demonstrating that they can and do practice forestry.[23]

Much of the argument in this volume would be revisited in *The People's Forests* and subsequent writings.

Following his work on the Copeland Report, Marshall was appointed Chief of Forestry for the Bureau of Indian Affairs, a position that brought him closer to the seats of Washington power and gained him access to figures of influence. With the election in 1932 of Franklin Delano Roosevelt and the subsequent institution of national programs to address the human wreckage of the Depression, the possibilities for socialization of areas of the nation's economy appeared heightened. For Marshall this meant an opportunity to place his case directly before the new president. His efforts were facilitated by Gifford Pinchot, then governor of Pennsylvania, who solicited from Marshall a statement to be presented to Roosevelt.[24] In that paper Marshall was clear about the program he envisioned: "The solution of the private forestry problem lies in *large scale public acquisition of forest lands, and the use of the unemployed to safeguard these public forests and help build up their productivity*" [his emphasis].[25] He also saw his plan as a partial solution to the problem of unemployment, putting the unemployed to work in the forests.

The unlikely possibilities of effecting such plans, acknowledged by Marshall in *The Social Management*, did not daunt his enthusiasm or his commitment to an advocacy for his vision. In 1933 *The People's Forests* was published and the arguments of earlier writings were expanded and augmented with facts and figures from the Copeland Report. The book is polemical and passionate; the struggle to save American forests demanded nothing less. Response to the book was predictable. The progressive press lauded it, finding an argument to bolster their hopes for expanding Roosevelt-era social programs. William Akers, writing for the *Federated Press Washington Weekly Letter,* noted: "The 'national plan for American forestry' proposed in the Copeland committee report is to be pushed at this session of Congress and the lumber barons are preparing themselves for a battle." He continued: "There is going to be a fight over the people's forests, and Marshall's book is the handiest and most reliable source for the facts."[26] Other laudatory comments appeared in the *New York Times,* which called it "richly informative, well thought out, soundly backed by scientific and economic knowledge," and in the *Nation,* which claimed it "should be a compulsory text in every American public school." Frank Reed, editor of the *Journal of Forestry,* expressed the contrary view: "This *People's Forests* is indeed a dangerous book. It is so attractively gotten up, so pleasingly written, and pleads its cause so plausibly that the superficially informed layman who peruses it is too certain to be artfully seduced to the support of a pernicious and subversive doctrine."[27] Despite lukewarm enthusiasm from official circles for Marshall's socialization program—Senator Copeland praised the report[28] but expressed misgivings about Marshall's plan as too ambitious and comprehensive—Marshall continued to argue the case. An article in the *New Republic* in 1934 spoke of the "evils of private ownership" and stated the urgency he felt: "If we want to preserve our forests and all that they mean to us, we must immediately undertake the burden of transferring them from the hands of private devastators to those of public rehabilitators."[29]

Bob Marshall's vision of a transformed structure of management for American forests was not realized in the Roosevelt programs. As Glover notes, the recommendations of the Copeland Report were not translated into law, at least not the whole package.[30] Marshall himself may have gained a sense of how difficult the task would be when he participated in a high-level conference on forest conservation in October 1933. Serving on a panel on "Public Timber Disposal and Public Acquisition," Marshall found himself outflanked and outvoted by delegates who represented the timber industries and who, despite endorsing the Copeland Report acquisition proposals, rejected the call for public acquisition and control proposed by Marshall.[31] A reading of the proceedings of the conference suggests that he was forceful, iconoclastic, and ultimately unsuccessful.

Although the socialization program envisioned by Marshall was not to be, it is clear that he saw an enhanced role for the federal government as the way to achieve his other great goal, the preservation of wild areas.

Wilderness and Activism

In the last five years of his life Bob Marshall pursued his goal of wilderness preservation from both his position inside government and as a private citizen outside this framework. He had written in 1930 that wilderness protection would require collective effort among like-minded advocates, noting that there were

> only a few years until the last escape from society will be barricaded. If that day arrives there will be countless souls born to live in strangulation, countless human beings who will be crushed under the artifices raised by man. There is just one hope of repulsing the tyrannical ambition of civilization to conquer every niche on the whole earth. The hope is the organization of spirited people who will fight for the freedom of the wilderness.[32]

A few years later he returned to this theme:

> Therefore, the fate of unmodified Nature rests in the activity of its friends. If they continue to be too busy or too indifferent to unite in its defence, then the universe of the wilderness is doomed to early extinction. If, on the other hand, they believe that its preservation is worth the sacrifice of some precious time and energy, and if they will take the trouble to become vociferous, there is no reason why material areas of America should not be kept primitive forever.[33]

It was an admonition that he took seriously. In August 1934 he met with Benton MacKaye and Harvey Broome in Tennessee to begin a conversation that would result in the formation of the Wilderness Society. Their meeting led to others, and in October a letter of invitation was sent out which made the case for their collective vision. At a meeting in January in Washington, D.C., in which Robert Sterling Yard and Harold Anderson were included, the designation of the Wilderness Society was adopted.[34] An invitation to serve in the formation of the society resulted in the addition of Bernard Frank, Aldo Leopold, and Ernest Oberholtzer to the organizing committee. The original statement of the launching of the Wilderness Society begins with words that Bob Marshall might have penned:

> Primitive America is vanishing with appalling rapidity. Scarcely a month passes in which some highway does not invade an area which since the beginning of time has known only natural modes of travel; or some last remaining virgin timber tract is not shattered by the construction of an irrigation project into an expanding and contracting mud flat; or some quiet glade hitherto disturbed only by birds and insects and wind in the trees, does not bark out the merits of "Crazy Water Crystals" and the mushiness of "Cocktails for Two." Such invasions are progressing everywhere so rapidly that unless they are fought as ardently as they are pressed there will soon be nothing left of those wilderness characteristics which make undisturbed nature the most glorious experience in the world to so many people.[35]

The statement goes on to detail the platform of the society and to revisit Marshall's classification of wilderness types and their depredations. It concludes with an appeal for contributions, noting that "already we have received one $1,000 anonymous contribution."[36] The initial contribution was from Bob Marshall, a commitment that he maintained throughout the early days of the society.

In developing the idea of the Wilderness Society, Marshall succeeded in convincing others that the movement had to be comprehensive—that the scope of the enterprise had to be the whole country, not just some regional or local focus. This was the translation of the understanding he had come to in Alaska that protection of wilderness must encompass whole systems. Another of his arguments, first articulated in 1928, concerned who would benefit from the protection of wild areas.[37] He was fully aware that a relative few could ever have the opportunities he had experienced to explore these places and appreciate the solitude they contained. Stating this as a "minority right," an opportunity denied to all if the despoiling continued, he made a case for the perpetual availability and presence of wild areas. James Wilson, Theodore Roosevelt's Secretary of Agriculture and the first to administer the Forest Service, had directed the management of public resources "for the greatest good of the greatest number," a concept often employed to counter Marshall's vision. However, those who invoked Wilson conveniently excised the rest of the quote: "in the long run." Clearly Bob Marshall took the long view, that the value of wilderness lay in its permanence, a resource for generations to come.[38]

In the years that followed the founding of the society Bob Marshall lent his considerable energies to the protection of numerous wild areas. He joined Ernest Oberholtzer and Sigurd Olson in their efforts to preserve the pristine nature of the Quetico-Superior area that spans the boundary between northern Minnesota and Canada. That struggle resulted in the designation in 1938 of a roadless area in the

Superior National Forest. The establishment of Voyageurs National Park in 1971 and the Boundary Waters Canoe Area in 1978 would provide later protection. In another part of the country he took up the cause of wilderness preservation on Washington's Olympic Peninsula. There the threat to a magnificent old-growth rainforest was acute. The salvation came in the creation of Olympic National Park in 1938, with 95 percent of its area becoming protected wilderness.

In Bob Marshall's work within government he continued to push for forest acquisition and the protection of undisturbed areas. He had become chief of the Division of Recreation and Lands in the Forest Service in 1937, a position that gave him more access to those who would have to create the safeguards he sought. Through his recommendation 4.8 million acres of Indian reservation lands were preserved as roadless. And in 1939, just two months before his untimely death, Marshall saw the adoption by the Forest Service of one of the most significant measures for preservation and protection. The U-Regulations designated wild and wilderness areas, establishing unprecedented conservation regulations governing these.

Bob Marshall's accomplishments were many and important; his work has been carried on by many in the Wilderness Society and by advocates in and out of government. The passage of the Wilderness Act of 1964 is a lasting legacy to his work, and its opening statement defining wilderness, while written by Howard Zahniser, could have been written by him: "A wilderness, in contrast with those areas where man and his own works dominate the landscape, is

hereby recognized as an area where the earth and its community of life are untrammelled by man, where man himself is a visitor who does not remain."[39] In 1980 the creation of the Arctic National Wildlife Refuge closed off to
most development a vast area of Bob Marshall's beloved
Alaskan wilderness. He did not live long enough to see the
enactment of these measures, but they were surely part of
his encompassing vision.

In *The People's Forests*, Bob Marshall pauses in his argument to remind us of the reality of what might yet be saved
and of his love for it.

> In the Selkirk Mountains of northern Idaho are many pri-
> mordial forests of towering white pines and Douglas firs
> which overshadow an understory of densely crowned cedars
> and hemlocks and Englemann spruces. As a person walks
> among the disordered trees and gazes upward through the
> many tiers of the dark canopy to where it brushes against
> the sky, he has the impression of a perpetual twilight and
> of almost infinite depth. He sees the most amazing variety
> around him: variety of tree trunk, from the light gray,
> deeply grooved, clean-boled shafts of the white pine, to
> the reddish, flakey-barked trunks of the hemlocks, with
> thickset branches almost to the ground; varieties of needle
> patterns, from the long-leaved whorls of the white pine to
> the light green rosettes of the larches and the dark, bristly
> spikes of the spruces; variety of size, etc. . . . The wind blows
> freshly, mysteriously, guardedly, through these great for-
> ests, and with it comes a feeling both of peace and of infi-
> nite complexity.[40]

What more might we say of Bob Marshall, of a life lived with such enthusiasm, of a sense of purpose pursued with such dedication? We ought to take note of the qualities elaborated in John J. Mitchell's tribute—Bob Marshall's humor and his humanity.[41] The wonderful quality of his life is that he loved people. His advocacy for wild places is always couched in terms of the human benefits of being in these places, the aesthetic, the invigoration, the escape from the humdrum and debasement. He was no isolated druid. He loved parties, he loved dancing, he was a movie buff, he was nuts about baseball, and he enjoyed good arguments and good friends. In short, he was the antithesis of the environmentalist misanthrope. He was, quite simply, a man who embraced life and lived it passionately. To our lasting benefit one of his great passions was the outdoors and wild areas.

I wish I could have met him.

Notes

In the preparation of this introduction I have been greatly aided by a number of works. James M. Glover's biography, *A Wilderness Original: The Life of Bob Marshall* (Seattle: The Mountaineers, 1986), is an indispensable source. A useful sketch of Bob Marshall's life and work is found in Roderick Nash, *Wilderness and the American Mind*, 3d ed. (New Haven: Yale University Press, 1982): 200–209, 287–89. There are two recent pieces on Bob Marshall: Donald Dale Jackson, "Just Plain Bob Was the Best Friend Wilderness Ever Had," *Smithsonian* 25 (August 1994): 92–100; and the luminous tribute by John G. Mitchell, "In Wildness Was the Perception of a Smile: An Evocation of Robert Marshall," *Wilderness* 109 (Summer 1985): 10–21. I am also indebted to David

Farrell for his help in gaining access to the Robert Marshall Papers at the Bancroft Library at the University of California, Berkeley. Assistance from Chad P. Dawson and James M. Glover also enabled this project to come to fruition.

1. Robert Marshall, *Arctic Village* (New York: The Literary Guild, 1933): 9.

2. Robert Marshall, "Wilderness as a Minority Right," *U.S. Forest Service, Northern Region, Bulletin* 12: 5–6 (1928), and "The Problem of the Wilderness," *Scientific Monthly* (February 1930): 43–52.

3. Glover, 12.

4. Ibid., 12–14; Andrew Johnson, "Louis Marshall, Adirondack Wilderness and the College of Forestry," in *Bob Marshall: Essays on the Life of America's Premier Wilderness Advocate* (Syracuse: State University of New York, College of Environmental Science and Forestry, 2000): 11–12.

5. Chad P. Dawson and Ed Zahniser, "The Influence of the Adirondacks on the Wilderness Preservation Contributions of Robert Marshall and Howard Zahniser," *USDA Forest Service Proceedings* RMRS-P-15, vol. 2 (2000): 45–48.

6. Glover, 67–97.

7. Marshall, *Arctic Village*.

8. *A National Plan for American Forestry* (Copeland Report), Senate Document 12, 73d Congress I, March 13, 1933.

9. Robert Marshall, *The Social Management of American Forests* (New York: League for Industrial Democracy, 1930).

10. *A National Plan*, 465.

11. Ibid., 465–66.

12. Marshall, "Problem of the Wilderness," 143.

13. *Social Management of American Forests*, 9.

14. *A National Plan*, 470.

15. *Arctic Village*, 50, 60, 372.

16. Ibid., 103.

17. Ibid., 112.

18. Ibid., 198.

19. Ibid., 113.

20. Ibid., 365–72.

21. Nash, 288.

22. *The Social Management of American Forests*, 25.

23. Ibid., 34.

24. "A Forest Program," manuscript in Robert Marshall Papers, Bancroft Library (RMPB), University of California, Berkeley.

25. Ibid., 3.

26. Akers, "Marshall Warns of Peril to U.S. Forests," RMPB.

27. The three quotes are reproduced in a magazine advertisement for the book, RMPB.

28. In a memo Copeland expressed his view that Marshall's chapter in the report was "as fine a statement of the place of the forest in recreation as I have ever seen," RMPB.

29. "Public Forestry or Private Devastation?" *New Republic* 79 (1021): 177.

30. Glover, 148.

31. First Conservation Conference, October 1933, RMPB.

32. Marshall, "The Problem of the Wilderness," 148.

33. Marshall, "The Universe of the Wilderness," 240.

34. Glover, 180–81.

35. "The Wilderness Society," 1, RMPB.

36. Ibid., 4.

37. Marshall, "Wilderness as a Minority Right."

38. Wilson also wrote in a letter to Gifford Pinchot, the first Chief Forester: "In the administration of the forest reserves it must be clearly borne in mind that all land is to be devoted to its most productive use for the permanent good of the whole people, and not for the temporary benefit of individuals or companies." See http://www.lib.duke.edu/forest/usfscoll/namechng.html.

39. Wilderness Act of September 3, 1964 (P.L. 88–577).

40. *The People's Forests*, 189–90.

41. Mitchell, "Perception of a Smile," 10.

TO

GEORGE P. AHERN
EARLE H. CLAPP
EDWARD N. MUNNS
GIFFORD PINCHOT
RAPHAEL ZON

Courageous foresters who for years

have been battling effectively and

uncompromisingly for

the social management of

American forests.

Acknowledgment

IF IT is legitimate to prophesy in a note of acknowledgment, I would like to venture the prediction that when the history of American forestry is written, 1933 will stand out not as the year of the Civilian Conservation Corps, or of huge public works appropriations or of a nationwide lumber code, but as the year when the United States Forest Service published *A National Plan for American Forestry*. This outstanding government document not only contains by far the most detailed and accurate statistics ever gathered on American forest conditions, but also advances the most progressive

Acknowledgment

interpretations. I have acknowledged by foot-note many specific data taken from this publication, but I also want to stress in a more general way its basic importance to me in the preparation of this book.

To Edward N. Munns I am deeply indebted for his cheerful willingness to discuss any points which perplexed me while I was in the process of writing this book and who criticized in detail my chapter on soil and water conservation, a subject in which he is an outstanding expert. Raphael Zon, Edward I. Kotok, and Stuart B. Show also helped me by especially stimulating discussion. Elise Untermyer was kind enough to go over each chapter as I was rewriting it and to make invaluable suggestions for improving the English and the method of presentation. George Marshall reviewed the entire manuscript in its semi-final form and made many detailed and exceptionally acute criticisms. To Burton E. Livingston I want to express deep appreciation not only for revising the physiology and grammar of my biological chapter, but also for continually stressing during two years of study under him the necessity of saying exactly what one means in technical writing.

Acknowledgment

Robert Sterling Yard, ardent fighter for the preservation of natural conditions in the outdoors, contributed valuable suggestions for my two recreational chapters. William N. Sparhawk was kind enough to review my chapter on government regulation of forests. My ideas concerning the relative importance of government regulation and government ownership were notably clarified by an exceptionally stimulating argument with Edward C. M. Richards. L. F. Kneipp and Abram L. Harris gave me very helpful criticism on the chapter dealing with method of acquisition, the former curbing considerable wishful thinking on my part, and the latter preventing several appalling economic fallacies. Henry B. Steer gave me essential data for my valuations of the different sorts of forest land.

Finally, I am obligated to Nan Dowling, Lilian W. Abercrombie, and Ella Jay for their intelligence and patience in recording the original draught of this manuscript as I dictated it to them.

ROBERT MARSHALL

Klamath Falls, Oregon,
September 13, 1933

CHAPTER I

Devastation Sweeping
the Country

W HEN the Pilgrims landed on the "bleak
New England shore" there spread before them
a verdantly covered continent of natural vegetation—an immense expanse of green forest and
green grass lands, of fresh flowers and luxuriant
herbage, of deep, black soil rich in all the organic
elements which are the basis of fertility.

At that time approximately half of what is now
the United States was covered with timber. Some
of this was open or scrubby growth, on the edges

[13]

of deserts or near the summits of mountains, but for the most part the woods exhibited all the features generally associated with the virgin forest. There were unbroken tracts embracing millions of acres which were covered with giant trees towering 100, 200, or even 300 feet into the air. Far below, on the floor of the forest, was a constant, mysterious duskiness where the succulent wood sorrel, twinflower, cornel, and goldthread blossomed with a freshness only possible in untrodden spaces. Among the dark green leaves of the flowers was a soft, resilient cover of moss in which the feet of the wild animals and the occasional Indians settled deeply and gently. There were no noises except the noises of nature, no scars except the scars of nature, no sights except the sights which nature had been exhibiting for countless centuries.

Of course, nature then as now was not static. Obviously, there had once been a time when there was no life on the continent. There had been incalculable ages after the dawn of life before the first tree had been evolved, and after the first primitive arboreal forms it had taken countless ages to develop forests resembling those of the present. During this almost imperceptibly slow development the entire forest had been wiped out more than once by ice sheets sweeping down from the north.

All the while the unceasing process of erosion was going on so slowly that to the Indians then living on the continent it was of no moment whatsoever. These native inhabitants, numbering probably no more than three-quarters of a million people, were so sparsely distributed over the nearly two billion acres of land in the present United States that they did not by their mere numbers hasten nature's process of erosion to any appreciable extent. Furthermore, they had for the most part an attitude of preservation, a realization that their life came from a nature which it would be catastrophic to destroy, and so with few exceptions the land in which they had been living for generations was left by them as rich and productive as when their ancient Mongoloid ancestors had first migrated to the North American Continent.

Nobody knows for how many centuries the Indians lived thus in equilibrium with their natural surroundings. However, the story of what has happened during the 300 years since the white race came to wrest the continent from its aboriginal occupants is patent to anyone who cares to read from nature. During the first century of settlement the pioneers were still so few in number that in spite of their destructive habits they were able to do only insignificant damage to the forest resources of the country. True, in order to obtain lye, they burnt down magnificent forests which had taken centuries in their development; burnt entire mountain ranges because they were too indifferent to control the fires which they merely needed for clearing up tiny garden patches; allowed their camp fires to run wild because they were too lazy to pour a little water on them when they departed; and slashed down giant trees simply to obtain bark for tanning extract, boughs for bedding, or even gum for chewing. The instinct for destruction was present from the earliest settlement, but the number of people to effect this destruction was so small and the forest resources so immense that for a long time the damage went practically unnoticed.

Yet this damage continued year after year, generation after generation, century after century, and was constantly multiplied as the population of the country multiplied. Eventually, around the close of the Civil War, a few perspicacious people began to notice that something disastrous was in the offing for the American forests. When these clear-sighted idealists pointed out what was happening to the national timber resources, the practical men of those days hooted them into oblivion. "Our forests disappearing?" they guffawed. "Why, that's the most nonsensical notion ever suggested! And besides, supposing they are, what can you do about it?" When the visionaries of those days, the Franklin D. Houghs, the Carl Schurzes, and the Verplanck Colvins, replied that we must set aside huge areas of public forests, must regulate the exploitation of private lands, and must protect the forests in general from the ravages of fire, insects, and fungi, they were promptly dismissed as giddy children who scorned the realities of life.

Meanwhile the course of destruction continued. Until 1850 Maine had been the chief lumber state in the Union. Because of the relatively light demands for timber it had been supplying the markets of the Atlantic Seaboard for more than 200 years without any terrific drain on its resources, but by the middle of the nineteenth century it was beginning to exhibit the result of

two centuries of destructive exploitation. About
this time New York came to the forefront among
lumber-producing states, and between 1850 and
1860 more timber was cut within its borders
than in any other state in the Union. During
the next decade Pennsylvania took the lead, with
white pine still the tree chiefly desired. About
1870, Michigan, Wisconsin, and Minnesota commenced a 30 year leadership in American lumber
production. By the early part of the Twentieth
Century they had been so thoroughly logged,
burned, and cleared that any adequate tree
growth was precluded over much of their area
for generations into the future. Before these
Lake States were cut out the center of the lumber
industry shifted to the pineries of the South.
Though natural conditions were more favorable
to tree growth than in almost any section of the
country, destruction was even more intense than
in the regions of past devastation. Fires, the
greatest of all forest enemies, were more prevalent in the South than in any other region
which had ever been logged, while the improved
markets made the lumbermen leave fewer acci-
dental seed trees than they had done in the earlier
operations when many trees, too small to be cut,

had unwittingly been the source of reproduction for a new forest. As the South definitely commenced to go downhill in production after 1916, the West Coast increased its lumber cut steadily until at the outbreak of the depression of 1929 the three states of Washington, Oregon, and California were contributing almost as great a volume to the nation's lumber supply as did the entire South. On the West Coast the huge trees were harvested by high-powered machinery, which resulted in a denudation compared with which the earlier horse or mule logging of the rest of the country even at its worst seemed like splendid forestry.

And so we find that 300 years of American history have spread increasingly destructive logging throughout almost the entire United States. Devastation is sweeping the country from the Atlantic to the Pacific seaboard and from the Canadian Border to the Keys of Florida. The glorious stands* of virgin timber, unprecedented anywhere in the world for volume and beauty, have in most cases given place at best to inferior second growth forests, at best next to worthless,

* The term *stand,* as used in forestry, is practically synonymous with forest.

unsightly brush which at least holds the soil in place, and at all too frequent worst to bare rocks from which all soil and vegetation have vanished.

For nearly 60 years with increasing volume we have been shouting that we ought to preserve our forests. For nearly 30 years we have had a federal Forest Service which actually has been managing the comparatively small area of our government forests in a splendid manner. Nevertheless, in spite of all our conservation talk and in spite of our demonstration of what can be done, the condition of our forests as a whole has been growing steadily worse. The reason has been that for most of our forest area we have only been willing to talk conservation. When it came to deeds, the sole action which we were willing to take involved such steps as seemed easy to accomplish. At last the time has come when we cannot help realizing that talk and easy steps will not even stop the increase of devastation, let alone build up our forests as they must be built up if we are to restore them to anything approaching their original productivity.

It is like a man who has been told by the doctors that he has a cancer. At first it is small, and a relatively minor operation will correct his illness. But he is frightened by the necessary sacrifice he must make for even the minor operation, and he is skeptical of the prognosis, so he takes some patent medicine which the quacks assure him will cure his trouble without the necessity for any serious pain. When at last it becomes apparent that this cure is a fake he again listens to the doctors, who tell him that a much larger operation is now necessary. This time he begins to believe them, but he still cannot summon the courage to tell the doctors to go ahead with something so vital and serious. He continues procrastinating and buying patent medicines which advertise preposterous promises, and all the while his cancer grows worse until it has developed so prodigiously that it is impossible to save him.

Of course, the death of the forest and the death of man are not quite the same. When a man dies it is the end. When a forest is completely devastated, when all the soil is washed off the mountainsides and nothing but bare rocks remains, it is still true that nature, given time, will build things up again. During the course of geological ages there will be further upthrusts and more rock disintegration which will restore the mountains and the soil which have been washed away. Geologists claim that upthrusts have occurred on the North American continent on the average of at least once every 400 million years, while soil experts say that it may only take a thousand years to build up an inch of soil. In much less than a billion years we may be again all that we were before the advent of the white race.

The only trouble is that the virgin forests of a billion years from now are of no more significance to us who inhabit the world today or tomorrow or a century from now than are the acacia or magnolia trees of Mars. Rocks and mountains may be ageless, but men and society are emphatically of the present, and they cannot wait for the slow process of nature to retrieve the catastrophe caused by their unthinking destructiveness.

CHAPTER II

Forest Values—Raw Material

THE first question which must be resolved is whether the devastation and deterioration of our forests is worth stopping. If we do not need our forests sufficiently badly perhaps we had better let destruction keep on at its accelerated pace rather than pay the cost of preventing it. Before reaching any decision it is important to examine the major values of our forests and consider whether they are worth the admittedly great price it will cost to retain them.

[23]

What are the major values of the forests? Fundamentally they may be divided into three categories. First, the forests are important as a source of raw material. Second, they are essential for the conservation of our soil and water resources. Third, they are for many people the most precious environment for recreation. Let us consider each of these values in detail.

The raw material obtained from the forests may be divided into two classes. The first class includes all wood products derived from the forest. The second includes the non-wood products, such as wild life, forage, and turpentine. Economically the former class of products is of much greater importance than the latter.

The volume of wood annually removed from the United States forests during the years immediately prior to the depression of 1929 is indicated in the following table: [1]

Product	Billion Cubic Ft.
Lumber	7.4
Fuelwood	4.0
Ties	0.6
Fence posts	0.6
Pulpwood	0.6

[1] Index numbers throughout refer to correspondingly numbered books in references at end of book.

Product	Billion Cubic Ft.
Cooperage	0.3
Mine timbers	0.2
Veneer logs	0.2
Logs and bolts	0.2
Shingles	0.1
Miscellaneous	0.3
Total	14.5

To express this figure in a more visual form, 14,500,000,000 cubic feet would make a solid block a mile square and almost as tall as the Washington Monument, that is, 520 feet high.

Lumber, it will be observed, embraces more than half of all the wood harvested from the forests of the country. It is utilized in tremendous quantity for houses, boxes, furniture, railroad cars, automobiles, woodenware, coffins, cabinets, agricultural implements, and hundreds of lesser products. It not only is volumetrically by far the most important forest product, but also makes the most fastidious demands on the quality of the timber from which it is manufactured. Merchantable lumber could hardly be sawed from a tree of less than six inches diameter, while it is seldom financially profitable to cut a tree for lumber when it is under 12 inches. One of the most important controls of lumber quality is freedom from knots. The smaller the tree, other things being equal, the knottier it is. The more widely spaced the trees are, the bigger and longer lived are their branches, and the larger and more numerous are their knots. Consequently, the best quality of lumber generally comes from good sized timber which at least in its youth has grown in a fairly dense forest.

The number of years that it takes to raise merchantable lumber depends on many factors. In the region of fastest growing timber, the Southeast, forests capable of producing merchantable lumber can be grown in 40 years. On the other hand, in many of the lodgepole pine stands of the Rocky Mountains it takes at least 150 years to grow forests capable of producing lumber. These figures, however, vary greatly, depending on the soil, rainfall, topography, drainage, and species of trees. Thus we cannot generalize that trees in the Southern States grow three or four times faster than those in the Rocky Mountain region. There are many white pine stands in Idaho which grow a great deal faster than the black gum forests of Alabama. Even when two stands are growing at exactly the same rate, some species of trees can be sold when they are only six inches while others may have such an intrinsically low quality of wood that they may not be marketable even when they are 20 inches in diameter. Again merchantable lumber depends a great deal on how close a tree is to the market. In some of the more remote stands in the Cascades and the Rocky Mountains, trees which might produce splendid lumber could not possibly be harvested because they are so inaccessible that the cost of getting them to the mill would be prohibitive. At the same time far inferior trees located within a few miles of a sawmill may be readily merchantable.

The problem of meeting our fuelwood demands is much simpler. Almost any quality of wood can be used for fuel provided it is not excessively rotten. In general fuel may be grown and harvested as a by-product of other forest resources. Trees which are of inferior species or condition and will not therefore qualify for the more valuable products, may readily be converted into fuel. Many farmers are able to eke out their slender resources by cutting fuel in their woodlots and selling it locally. Foresters have discovered that they can often pay for desirable cultural operations in the forest, such as thinning out too dense stands and removing dead and defective trees, if they can sell what they remove for fuel.

Railroad ties consume the third greatest volume of wood. The quality of wood demanded for ties is much higher than for fuel. Formerly only a few species of trees were considered sufficiently durable to withstand the combined ravages of the mechanical pressure of the passing trains and the fungi which rotted the ties away. Within the last two decades, however, tremendous progress has been made in wood preservation. By treating ties with creosote or other chemicals, fungi are prevented from developing, and thus the average life of ties has been increased two to tenfold. Many species of wood which could not possibly be used before the era of preservatives, now are found to make excellent ties.

The quantity of wood required for fence posts is easily obtained, since posts, like fuel, may often be cut as by-products in the harvesting of lumber. In the past the chief quality of fence posts has been that they must be durable in contact with the ground. Today that portion of the fence post which is in or near the ground is often treated with creosote or other preservatives, and so durability is no longer as important a requirement as it formerly was.

Pulpwood, while it only ranks fifth in volume of wood removed from our forests, is actually next to lumber the forest product which is most difficult to obtain. This is chiefly because only certain species of wood have sufficiently long fibers to make satisfactory pulp. Until recently the pines, even though they had long enough fibers, contained too much resin to be utilized. However, recent chemical research has made it possible to use this rapidly growing genus. New discoveries have also made available various dark fibered woods, such as hemlock, which formerly were unsatisfactory for most pulp products. Nevertheless, spruce still constitutes 70 per cent of all the pulpwood which we consume in this country,[2] although hemlock and the pines are beginning to give it good competition.

Even before the passage of the 3.2 per cent Beer Act, cooperage stock for making barrels ranked sixth in quantity of wood consumed annually. Cooperage requires hard wood and in many cases wood which has no flavor to taint the product contained in the barrel. Cooperage is of two types: slack in which staves do not fit tightly together; and tight in which liquids can be stored without leakage. The chief cooperage woods are the northern hardwoods, beech, birch, and maple. The removal of these trees is often silviculturally desirable when logging in the spruce-hardwood forests of the Northeast, and the demand for them may actually result in the growing of more rather than less timber.

Mine timbers constitute a relatively minor drain on the forests. The demand for them has been small, but fairly constant over a fairly long period of years. They are used for bracing shafts and galleries, and their chief requirements are strength and durability.

The demand for veneer logs, which was practically non-existent prior to 1905, nearly doubled in the decade from 1919 through 1929.[2] Veneer is used a great deal for modern furniture, thin slices of choice wood being taken to cover a core of less valuable lumber. Plywood, consisting of thin strips of wood pasted together, is a recent veneer product remarkable for its strength. The label on the cover of this book is an example of a minor veneer use.

There is also a small consumption annually of logs and bolts which are used directly in the manufacture of such articles as vehicle stock, handles, woodenware, and furniture. These logs and bolts generally require a relatively high quality of timber.

Composition roofing has cut the use of shingles to less than half of what it was in 1910.[2] Many cities have laws prohibiting the use of shingles for roofing because of the fire hazard involved. While shingles are unquestionably more inflammable than high grade composition roofings, they are on the average fully as durable, and furthermore they have an insulating value which none of the composition roofings can approach. The one major wood used for shingles is cedar.

It is important to emphasize that all of these major forest products except ties, mine timbers, and veneer logs may be harvested chiefly as by-products in the production of lumber. Their development, instead of being a drain on the forest, could often under proper management be made an aid to the practice of good silviculture. (*Silviculture* embraces all activities in the woods which are conscious measures to maintain or improve the productivity of the forest.) Rotten and unmerchantable hemlock on the West Coast, which has frequently dominated the forest after the Douglas fir and spruce were logged, could be harvested for pulpwood. This would free the land from the encumbrance of the old, unmerchantable timber, and would make possible the production of a new, valuable, and rapidly growing forest. Pine stands in the Southeast, which often grow so densely that they nearly stagnate, could have large quantities of fence posts cut from them, and at the same time have their growth rate much accelerated through this thinning. Unmerchantable hardwoods, which often preclude good forestry practice in the spruce-hardwood forests of the Northeast because they prevent the softwood reproduction from developing, could be removed and converted into cooperage stock if barrel factories were established at places accessible to logging operations. Today certain companies are actually girdling their hardwoods after they have removed the spruce and hemlock, simply because without killing these large, unmerchantable trees it is virtually impossible for a new forest to get started. These are merely examples of how under proper management most of the minor forest products might be harvested with little additional drain on the timber resources.

There is, however, a very serious additional drain on the forest which has not yet been discussed. This drain involves nearly two billion cubic feet of timber which is annually destroyed by fire, insects, or disease. The average amount of wood burnt annually between 1925 and 1929 amounted to about 870 million cubic feet per year, while that destroyed by disease, insects, wind, drought, etc., and not salvaged, amounted to 940 million cubic feet. Thus the total drain on the wood resources of the United States, including both what is utilized and what is not, approximates 16½ billion cubic feet.[1]

In addition to wood products, there are a large variety of other forest commodities. Forage, the most important of these, is grazed upon by domestic live stock on some 334 million acres of forest and brush land in the United States. The animals which graze in the forest include sheep, cattle, hogs, horses, burros, mules, and goats. "Nearly 12 million grown sheep, largely ewes, which had in 1931 a value of about $60,000,000, are estimated as grazing on forest ranges", according to Chapline.[3] He also states: "It is estimated that during 5 to 8 months of the spring, summer, and fall these Western forest lands furnish feed for over 2½ million mature cattle, having a value in 1931 when prices were low of about $85,000,000." In many regions grazing is a serious detriment to the forest, for the animals tend to trample or eat the young trees. Too heavy grazing is especially conducive to erosion. However, with proper regulation there is no reason why grazing should not be a very important and valuable by-product of the forest.

Fish and game, quite apart from their recreational value, have considerable importance as a source of food. Many dwellers in or at the edge of the forest utilize the wild life either legally or illegally as an important food supply. Roberts states that the value of meat and fur taken annually amounts to approximately 190 million dollars and that of fish to 14 million dollars. The major share of this huge total is derived from forest lands.[4]

A very important forest industry is that of naval stores. In 1929 there were 36 million gallons of turpentine and 2½ million barrels of rosin recovered from the forests of the United States.[2] Naval stores are obtained by making light, narrow chippings on the sides of trees rich in resin, allowing them to bleed, and collecting this resin from the wound. The modern technique has greatly increased the life of resin trees and the total volume of naval stores obtained per tree, and has decreased the fire damage and the general destruction of the forest. Southern pines yield an overwhelming part of the turpentine and rosin harvested from the American forest.

Other non-woody forest products which have a definite commercial value include herbs, nuts, berries, flowers, bark for tanning extract and medicines, maple sugar, Christmas trees, holly greens, and fiber producing plants. These are not of major importance, but they are a part of the forest wealth.[5]

These varied materials derived from the forest indicate its tremendous value in the present American economy. The question naturally arises whether the demand for these products, and especially for wood, may be expected to continue in the future. There has been a great decrease in the amount of wood taken from the forests since the outbreak of the depression. Does this represent a permanent tendency, or may we count on the drain returning to normal when and if the depression abates?

An important indication that America's wood consumption may be reduced is that the demand for lumber, the major forest product, has decreased slowly during most of the past 25 years and tremendously since 1929. That has been partly because the rate of increase in wooden houses has abated since the frontier closed, and partly because wood has been displaced by the less inflammable building materials in city construction. Large cities often bar the construction of wooden houses, and buildings are being put up in many places without any wood even in floors and interior finishing. It is, therefore, natural that the per capita wood consumption in this country has dropped at a steady rate from 495 feet board measure* in the decade from 1900-1909 to 315 feet board measure in the decade of 1920-1929. The actual total lumber production in the country has taken the fol-

* A *foot board measure* is one foot square and one inch thick. It is the unit normally used for measuring lumber.

lowing course since the early part of the 19th
century: [2]

Year	Total Lumber Production Billion Board Feet	Year	Total Lumber Production Billion Board Feet
1809	0.4	1919	34.6
1819	0.6	1920	35.0
1829	0.8	1921	29.0
1839	1.6	1922	35.2
1849	5.4	1923	41.0
1859	8.0	1924	39.5
1869	12.8	1925	41.0
1879	18.1	1926	39.8
1889	27.0	1927	37.2
1899	35.1	1928	36.8
1904	43.0	1929	36.9
1906	46.0	1930	26.1
1909	44.5	1931	16.6
1914	40.5	1932	9.5

Many of the minor forest products tell the
same story. Fuelwood, ties, fence posts, cooperage
stock, and shingles have all declined in demand
during the past 20 years. The only ones of the
ten major forest products which have shown a
material increase have been pulpwood and veneer logs.[2]

Of the individual products which have

dropped, it is difficult to see how the demand for
lumber will ever come back to its former volume,
how fuelwood can ever occupy the position of
importance it once did before the development
of oil and electricity for heat and energy, how
the demand for ties can possibly do anything
but decrease with railroad expansion at a stand-
still and the life of ties greatly prolonged by improved methods of preservation, how fence posts
can regain the ground they occupied before
metal posts and wire rails came into use, or how
the consumption of shingles will ever again be
as great as it was in the days when composition
roofings were unknown. On the other hand
there are products which will probably show an
increased consumption of wood. Wooden furniture is more popular than ever. Cooperage
products, with the end of prohibition, should
certainly rise. Pulpwood has shown a consistent
increase for a century, and there is no reason
to suppose that this increase will not continue.
Furthermore, there are a great many entirely
new products which in recent years have been
developed from wood fibers, the most important
of which are rayon, cellophane, and various
fiber boards such as Masonite. These do not

make use of wood as such but of the cellulose which may be derived from wood.

This brings to mind the fact that wood uses are changing, and that the major wood demand tomorrow may not be for lumber, or fuel, or ties, or fence posts, or any product which uses the tree in its raw state. Indeed, it seems probable that our greatest interest will be in the chemical substance which goes to make up all species of wood, namely, cellulose. Cellulose is one of the most required chemical products in the world. The number of uses to which it is put increases annually by the hundreds. From every basis of intuition and reason it appears that the future demands for cellulose will be immense. Now, wood is not only the most compact form of cellulose known in nature, thus possessing a tremendous advantage when it comes to transportation, but in the well managed forest more cellulose per acre per year may be raised than with any other crop. Furthermore, it may in many cases be raised and harvested at virtually no cost as a by-product, in thinning out forests that need it badly or removing otherwise unmerchantable species in forests whose primary product may be the more valuable lumber or

ties. So we can safely say that the use of wood
for cellulose will be tremendous, and this may
largely offset the decrease in the use of other
forest products.

Another factor which may tend to offset the
present decrease in the demand for wood is the
softwood timber famine which is looming all
over the world.[5] Russia, the last great source of
coniferous timber, is cutting at a rate far greater
than her production, and the end of her readily
accessible timber supply is predicted within 35
years.[6] Canada's forests are following the same
course as those of the United States. France,
Germany, England, and Italy are all consuming
more lumber than they grow. Poland, Sweden,
and Finland, which grow more than they consume internally, cannot possibly supply the vast
demand of the neighboring countries. "Altogether," Zon and Sparhawk say in their scholarly
study, "a survey of world consumption and
world trade in wood shows no evidence of permanent diminution, except in the United
States." [6] With no decrease in worldwide demand, and with an international timber famine
growing more and more acute, the consequence
will be that any country with a supply larger

than is necessary to meet its own demands may reasonably look forward to an important export trade. The United States, because of its vast area and highly developed transportation system, is one of the few countries capable of growing more accessible wood than it will consume itself.

In considering the demand for timber it is fundamental to remember the demand for a product of any sort is never static. It varies with price, which in turn varies with available quantity. The low demand for platinum today does not mean that if platinum were very common and, therefore, relatively cheap, it would not be in immense demand. The case of aluminum is very instructive in this regard. In 1885 the cost of aluminum averaged around $16 a pound, and the amount of aluminum produced in the United States that year was only 283 pounds. Between 1886 and 1889 the electrolytic process for the production of aluminum was invented and developed. This was so much more efficient than the old process that by 1890 the price of aluminum had dropped to $1.00 a pound. Concomitantly the demand for it multiplied so greatly that in 1890 the total output of aluminum in the United States was 61,281 pounds, a more than two hundredfold increase in five years. While there had been hardly any demand for aluminum when it was scarce and expensive, a tremendous demand suddenly developed when it became abundant and cheap.

Exactly the same principle applies to timber. If we have enough forests to make timber readily available and cheap we will use a great deal of it. But if we continue to devastate our forests so that timber becomes a rare and expensive product, our standard of living will be reduced in accordance with the higher cost.

CHAPTER III

Forest Values: Soil and Water Conservation

OF ALL the dynamic forces of nature, the most relentless and persevering is the force of erosion. Continuously the power of flowing water is washing away the crust of the earth. Erosion is an endless and inevitable process, and it will never cease as long as any point of land juts above the level of the sea. Looking at the earth and its resources from a cosmic standpoint we can well ask: "What is the difference if our soil is being washed away and our lands flooded.

Such things have always been and always will be. We cannot change the laws of nature."

This may be a laudable attitude to take in the leisurely sophistication of an evening in a drawing room when there is no necessity to be concerned with the realities of the present. In a practical world, however, where a single rain storm may cause scores of deaths, ruin hundreds of families, and destroy the happiness of thousands of human beings, erosion and the floods that are associated with it are of vital significance.

Floods in the United States are causing havoc undreamed of by most people. The Mississippi River flood of 1927 alone inundated 18,000 square miles, drove 750,000 people from their homes, took 246 lives, and did approximately $300,000,000 worth of damage. Yet that was only one flood in one single year. There have been at least 16 floods since 1900 on river systems other than the Mississippi which have done more than $1,000,000 worth of damage.[7]

In the balance of nature there is a reciprocal reaction to floods which may be more disastrous than the first evil. Precipitation is limited. When most of the rain or the melting snow runs into the streams at one time, it follows that there will later be a shortage of water. That is why severe floods are commonly followed by severe droughts, with the ground turning hard as rock, the streams drying up, and the water level of reservoirs falling alarmingly.

The severity of erosion in the United States varies from a few pounds of dirt washed away annually on some exceptionally well protected acres, to many tons washed away on millions of acres which because of soil, topography, vegetation, and climate are peculiarly susceptible to erosion. "A single rain on experimental plots near Holly Springs, Mississippi, for example, removed soil at the rate of 23 tons per acre from cultivated land with a 10 per cent slope. Studies in southwestern Wisconsin have resulted in an estimate that an area of 10,000 square miles contributes 15 million tons of silt to the Mississippi River annually. In both sets of experiments erosion from forested soils was insignificant compared with that from barren or cultivated soils." [7]

By far the major portion of the billions of tons of soil washed away annually in this country is due to the mismanagement of man. A spring freshet which slightly muddies the water of a wilderness brook causes erosion which is unpreventable. But a spring freshet, rolling down a farm land furrow plowed against the contour of a slope, and washing out in a single season a gully 30 feet deep and 100 feet across, causes erosion which is entirely preventable. Furthermore this erosion proceeds at so rapid a pace that unless something is done to stop it the effects from many different standpoints will be disastrous.

The erosion problem in America embraces three classes of land. The first is the cultivated farmlands, the second the grazing and pasture lands, and the third the forests. While this book deals with the forest problems in America, erosion as a phase of the forest problems cannot be divorced from the erosion which occurs on non-forested lands.

H. H. Bennett of the U. S. Department of Agriculture, an expert on soil erosion, writes: "The annual cost of erosion in farm lands impoverished or ruined, highways damaged, reservoirs and irrigation ditches filled, and valley lands overflown is estimated at not less than $400,000,000. This does not take into consideration the accumulating damage that cannot be measured until the last inch of topsoil has been swept out of fields. At least 3,000,000,000 tons of soil are removed from the fields and pastures of the country every year. The potential value of the plant food contained in this material exceeds $2,000,000,000, on the basis of the cost of the cheapest commercial fertilizers." [8]

Bennett further states that 34,000,000 acres of formerly cultivated land have been destroyed for crop production. In addition 125,000,000 acres of land now in crops have been severely impoverished through the loss of all or most of the topsoil.[9]

The chief cause of erosion on farm lands is faulty cultivation. If a farmer plows his fields along the contours, or in other words plows horizontal furrows, much of the rain as it falls will collect in the furrows and seep into the ground. But if he plows the furrows against the contours, that is, up and down hill, the rain water that falls will find natural channels in the furrows and will flow along them as if they were brook beds. The loose soil on the top of the ground is very readily washed away, making a deeper brook bed which collects a larger volume of water. This flows along with greater and greater power, and its accumulating pressure keeps tearing ever deeper gashes into the ground. Other causes of erosion on cultivated land include the mere fact that any cultivation has taken place on steep hillsides with loose soils, the depletion of soil fertility and consequent decrease in permeability, the leaving of the ground without any crop to bind the soil during seasons of heavy rainfall, and the failure to check incipient gullies when they first become apparent.

It is perfectly possible to stop this erosion of cultivated land if the proper steps are taken *in time*. Bennett cites the case of two adjacent fields of Kafir corn in Kansas. In one the soil between the corn rows was bare, as it customarily is in cornfields. In the other, grass was sown between the rows. In the normal bare area 4,250 times as much soil washed away, and 399 times as much water ran off as in the grassed area.[9]

The methods of stopping erosion on cultivated lands may be classified as physical and biological. The most important physical steps include the substitution of contour plowing for cross contour plowing, the terracing of eroding hillsides, the sowing of cover crops at certain key points

which have the effect of inducing natural terracing, the digging of holes to collect the runoff
before it has a chance to flow very far, the throw-
ing of brush into gullies, and the building of
checkdams to prevent further washing. From
the biological standpoint the most important
step is to revegetate the eroded lands. This may
take the form of sowing grasses or any other
form of herbaceous plants to lock the soil with
their roots. It may involve maintaining cover
crops on the ground between the fall harvest
and the spring sowing of the commercial crop,
a time of year when the soil is normally bare.
It may mean that many lands, especially those
with a slope steeper than 15 percent, should be
taken out of cultivation altogether. It may also
involve the planting of trees, which are the most
effective controls of erosion and the best means
of building up the ruined fertility of the eroded
lands.

The problem of erosion in pasture or grass
lands is a much simpler one than that in cultivated land. The chief cause of erosion in these
areas is over-grazing. If animals browse in too
great numbers over any piece of ground they so
deplete the vegetative cover that the soil is no

longer bound in place by the roots of the vegetation. The scars made by the impact of animal hoofs, and particularly the creation of animal trails may also be responsible for starting channels of erosion. The most effective way of stopping erosion in pasture lands is to limit the grazing to an amount which will not deplete the vegetation or cause serious physical abrasion. Where soil washing has progressed sufficiently far, it may also be necessary to adopt the more expensive methods of control discussed for cultivated lands.

The chief causes of erosion in the forest lands are fire and grazing. Fire, by killing trees and herbaceous vegetation, removes the interlocking roots which bind the soil in place and also destroys the blanket of duff which acts as a cover for the mineral soil. Grazing in the forest has the same effect upon the ground cover as does grazing in pasture lands. Logging operations as such, even if they may be quite destructive to tree growth, do not generally cause severe erosion problems unless they are followed by fire. However, on very steep hillsides the mere physical stirring of the soil by the falling trees is often enough to induce serious soil washing. The ways to meet the problem of erosion in the forest are to stop fires, to regulate grazing when it is permitted at all, to conduct logging operations according to forestry principles, and to avoid logging altogether on especially steep hillsides.

Man by his activities has changed the forest from an entity in which soil building practically counteracted the influence of erosion to one in which the rate of soil destruction is immensely greater than soil creation. Good management could practically restore the former balance of nature. Without it we are doomed to that stupendous amount of erosion already mentioned, which clogs streams, irrigation ditches, river channels, harbors, and reservoirs, increases flood heights by filling up the bottoms of the rivers, and permanently ruins millions of acres of overflowed farm lands which become covered by deposits of unfertile, eroded soil.

Having considered very briefly the nature of erosion, let us examine in what way the forests affect streamflow. First, the soil under the forests consists of a layer of organic matter, often more than a foot deep, embracing several different strata. The stratum at the surface is composed of a litter of recently fallen leaves, twigs, and needles which have not yet lost their original appearance. Below the litter is a mould of the same material in a partly decomposed state, and below that a humus of black organic soil so decomposed that none of the original elements from which it was created can be distinguished. Below these organic layers is the mineral soil, into the upper portion of which some of the rich organic material has infiltrated, making it much more permeable than the pure, unorganic mineral soil. Together this organic layer of so-called duff and this enriched mineral soil act as a sponge in absorbing a vastly greater amount of water than is absorbed by the soil of open lands. Lowdermilk has recently shown that even more important than the sponge function of the duff is its blanket function. The mineral soil under the organic layer is filled with minute pores extending downward from the surface of the ground. It is these pores which retain most of the water held by the soil. The duff acts as a cover to prevent these pores from becoming clogged and thus permits them to absorb an immense volume of water. Thus the rain which falls in the forest, as well as the water from the melting snow, instead of immediately flowing down into the creeks and rivers is retained for varying periods, and only slowly released to the drainages of the region. In other words, the run-off from open lands is intense but of short duration, while that from forest lands is light but extends over a protracted time.

Second, the trees, by the shade which they cast, delay the melting of snow in the spring. Thus in forest lands the run-off from snow water is spread out by a protracted melting, while in the open the melting and the consequent run-off occurs in a much abbreviated period. It is further true that there is a smaller volume of snow under the forest than in the open, because the branches of the trees intercept a large amount of snow which is evaporated directly back into the air. This, of course, decreases the amount of flood water when the snow melts in the spring.

Third, trees draw from the soil an immense amount of water which is carried up through their trunks and is evaporated from their leaves by the process known as transpiration. The amount of water transpired into the air by a single birch tree on a dry day has been estimated to be as high as 100 gallons. With an entire forest of trees transpiring, the amount of water which never flows into the streams, but instead is returned directly to the atmosphere, can only be appreciated by people with vivid imaginations.

Finally, the forest, by helping to stop erosion, decreases the deposit of silt in river channels. It is amusing to consider that all the while that dikes of increasing height have been constructed along the channel of the Mississippi River, the river bottom itself has been filling in, and consequently the water surface has been forced upward at a material fraction of the upward progress of the rising dikes. If the present process continues for a sufficient time archeologists of the future may be astounded by a Mississippi River with a bottom at the level of the old river crest and two great dikes jutting far above the surrounding country.

Of course all forests do not have the same influence on stream-flow conservation any more than they do on soil preservation. The Forest Service estimates, however, that out of some 615 million acres of forest and brush land, 308 million acres have a major influence on watershed protection.[7] In other words, the maintenance of half the forest and brush land of the United States would be essential from a flood control standpoint if no other value were involved. Another quarter of the forest land has an important if not a dominant influence on flood control. The degree to which a forest influences the functioning of a watershed varies with the type of trees and underbrush, the character of the soil, the topography, and the climate. When all four unite in having a major influence, the effect of the forest on water and soil conservation may be so great that all its other values pale into insignificance.

CHAPTER IV

Forest Values: Recreation[10]

THE use of the forests for recreation probably dates from the time when some wandering savage, returning to his cave through the depths of the primeval forest, may have noticed a beam of sunlight shining on some darkened tree trunk and felt all at once without knowing why a moment of great, surging joy in the chaotic passage of his life.

Until very recently, the recreational enjoyment of the forest has been chiefly of this incidental nature. Today, however, forest recreation is no longer an unpremeditated matter. People do not, as a rule, live in the forest any more, and if they go there to enjoy themselves they are fully conscious of their purpose. Consequently they have come to realize that forest recreation has a definite value in their lives, that it is something for which they are willing to sacrifice time and money, and so they desire to plan for the preservation of its possibility.

The following table shows in round numbers the best available statistics on the use of American forest lands for recreation during the year

Type of land	Number of Visitors
National parks	3,000,000
National monuments	400,000
National forests	32,000,000
Other federal lands	1,500,000
State parks and forests	50,000,000
County and municipal parks	60,000,000
Private lands	100,000,000
Total	246,900,000

1931. These figures are valuable not as giving a precise measure, but for the general impression which they convey of the enormous popularity which recreational forest use has already attained in America.

There is, of course, a great deal of duplication in these records. Some people visited several different parks or forests; some people visited the same area on a number of occasions and were checked separately each time; perhaps some people made use of all seven types of land for recreation. No doubt there were individuals who were counted more than a score of times in the total figure. It is therefore impossible to state how many different persons made recreational use of the United States forests during the year.

It is reasonable, however, to estimate that each visitor spent on the average one full day in each park, forest, or private timberland for which he was recorded. Some remained only for a few hours, it is tue, but many remained several days, and a few spent the entire summer on a single forest area. If the estimate of one day per visitor is correct, a total of approximately 250 million man-days were spent during 1931 in recreational enjoyment of the forest.

The national parks and the national forests have kept count of the number of their recreational visitors since 1916 in the one case and since 1917 in the other. These records started just prior to the era when long-distance automobile travel became mechanically and financially possible for a large proportion of our population, when consequently old notions of distances were altered almost overnight, and when entirely new recreational habits were formed by millions of Americans. The astounding increase in the number of both park and forest visitors, as shown in the following table, reflects the fact that forest recreation has grown during the past 15 years from a relatively unimportant variety of diversion into one of the most universally adopted forms.

There is considerable duplication in the records presented in this table. The standards of tabulating visitors have also varied. Nevertheless, any resulting inaccuracy is not of serious

| Year | Number of visitors to | |
	National parks (a)	National forests
1916............	237,357	....
1917............	353,268	3,160,000
1918............	311,661	3,322,565
1919............	594,835	3,064,344
1920............	756,654	4,832,671
1921............	876,367	5,633,420
1922............	938,338	6,172,942
1923............	1,168,886	10,543,893
1924............	1,258,178	11,394,366
1925............	1,495,372	15,279,730
1926............	1,670,955	17,112,024
1927............	2,173,120	18,523,888
1928............	2,323,089	23,008,997
1929............	2,496,080	31,758,231
1930............	2,607,499	31,904,515
1931............	2,999,451	32,228,613

(a) Exclusive of the urban Hot Springs National Park.

moment in face of a 750 per cent increase in national park use and a 920 per cent increase in national forest use during the brief period of 14 years. There is no reason to suppose that the general trends on federal lands are different from those on other recreational territory. Consequently, it is fair to state that recreational use of forest lands is growing with tremendous acceleration.

Is there reason to believe that this present growth in recreational use will continue? What factors are likely to inhibit and to stimulate future demand for forest recreation?

The most serious of the possible inhibiting factors seem to be commercial exploitation and fires, which threaten to deplete severely the beauty of many recreational areas. There can be no doubt that the greatest attraction of the forests is their natural beauty. If this is not adequately safeguarded, unquestionably millions who now delight above all else in the loveliness of the forest will forsake it for some other source of recreation.

Another of the possible inhibiting factors may be readily observed at any popular camp ground over a holiday. Here one sees swarms of tourists who not only destroy, by their mere numbers, the very isolation which was one reason for their journey to the forest, but also kill the ground vegetation around the camp site and tramp down the soil so compactly that even the trees often die. This type of destruction has its remedy, like the destruction resulting from commercial exploitation; but unless the remedies are applied these factors will definitely tend to decrease the recreational use of the forests.

On the other hand, there are several causes which may be expected to stimulate the recreational use of our forests. First of all is the mere fact of our increasing population. The predictions of reputable biometricians place the eventual point of saturation in population for the United States between 145 million and 185 million, an increase over present population of between 20 and 50 per cent. This would probably result in a corresponding increase in the number of recreation seekers, even if no other factors were involved.

With the increase in population, there has come a general decrease in the hours of labor. Whereas the average full-time working week ranged from 40 to well over 60 hours in 1929, estimates for the not distant future place the average in some cases as low as 16 hours and in the majority of cases at least as low as 30 hours. The National Industrial Recovery Administration has already reduced the hours of labor to a maximum of 40 a week. It seems reasonable to assume that a fair share of the consequent increase in leisure will be devoted to forest recreation. However, if shorter working hours mean smaller income, the change may actually militate against this result.

The past two decades have increased immensely the ease of transportation to the forest. The advent of the cheap automobile has shortened distances and made it possible for people living 500 miles from attractive areas to visit them as readily and as inexpensively as people a few years ago visited places but 50 miles away. With airplane and autogyro transportation rapidly being perfected, there is every reason to believe that the forests will become still more accessible.

Last but by no means least, one psychological factor must be mentioned among the causes tending to increase the use of the woods for recreation. As society becomes more and more mechanized, it will be more and more difficult for many people to stand the nervous strain, the high pressure, and the drabness of their lives. To escape these abominations, constantly growing numbers will seek the primitive for the finest features of life.

An evaluation of forest recreation may be attempted upon either a monetary or a social basis. In terms of dollars and cents, there are two especially suggestive ways of estimating the recreational worth of the forest. One is to determine the number of people who visit the forest annually, ascribe some average value to the pleasure which each one gets from it, and multiply. The other is to calculate the amount of money which recreationists spend in visiting the forests.

It seems distinctly an understatement to hold that each all-day visitor to the forest derives as much pleasure from it as he would derive from a 2-hour motion-picture show. I have estimated that in the United States approximately 250 million man-days a year are devoted to forest recreation. If the admission price to a movie averages 25 cents, this gives the annual American forest recreation a value of $62,500,000. This is the minimum that people probably would pay for the privilege of using the forest if the price were asked. The incidental fact that people have to pay for admission to the movies and do not usually have to pay for admission to the forests does not mean that the outdoor recreation is any less valuable.

The American Automobile Association estimates that almost $4,000,000,000 was spent in motor camping and vacation travel in the United States during the year 1929. It does not seem unreasonable to assume that one-quarter of the vacation motor travel is through forested country, and this would mean that annual forest vacation motor expenditures amounted to about $1,000,000,000.

The Special Senate Committee on Conservation of Wild Life Resources calculates that in 1929 hunters and fishermen spent $650,000,000 in addition to transportation expenses.[11] It would be conservative to estimate that at least three-quarters of this enormous sum was spent on forest hunting or fishing, which would mean that this form of forest recreation accounts for an annual outlay of approximately half a billion dollars.

No national figures are available for the amount of money spent each year on summer homes, hotel and resort accommodations (other than those paid for by automobile tourists), hiking equipment, or the outfits required for wilderness journeys. These expenditures would

unquestionably run half as high as those for
hunting and fishing. Consequently the following
would seem to be a reasonable, although admittedly a very rough estimate of the amount of
money spent on forest recreation during the
peak recreation year of 1929:

Forest vacation motor travel	$1,000,000,000
Hunting and fishing	500,000,000
Summer homes, resorts, hiking, wilder- ness journeys	250,000,000
	————————
Total	$1,750,000,000

Although huge sums of money are involved
in any basis of calculation, the most important
values of forest recreation are not susceptible of
measurement in monetary terms. They are concerned with such intangible considerations as
inspiration, aesthetic enjoyment, and a gain in
understanding. It is no more valid to rate them
in terms of dollars and cents than it would be to
rate the worth of a telephone pole in terms of
the inspiration it gives. The only common de-
nominator for the recreational and commodity
value of the forest is the human happiness which
may be derived from each use. Unfortunately no
quantitative measure of human happiness has

ever been designed, and consequently it is impossible to estimate accurately the contribution which forest recreation makes to the welfare of mankind. About all one can do is to point out the purposes for which men seek the forest and let each reader make his own evaluation tion of their intrinsic importance.

A frequent purpose of those seeking forest recreation is simply to have a good time in the outdoors. A large fraction of the vacation motorists enjoy what features of the forest they can observe at a velocity of over 40 miles an hour, but never really transfer their lives from the highway to the forest. The woods are only a pleasant background for a type of satisfaction which could be enjoyed with no essential diminution in any open-air environment. The urge to go somewhere, and the desire for amusement and a change from everyday life are the vital concerns with this particular type of recreationist.

Millions of people go to the forest for the preservation, restoration, or stimulation of their health, and for the physical exhilaration which they derive. The therapeutic value of forests has long been recognized in selecting sites for sanatoria. More recently it has come to be real-

ized that healthy people may be immensely
benefited by the pure air, the joyful exercise,
and the peaceful relaxation which are such important attributes of forest recreation.

There are many people whose primary concern in the forest is the pursuit of beauty. The
joy which such people derive from a sunset
across some forest lake, from the boundless immensity of the panorama off some wilderness
summit, from the unmarred splendor of a virgin
forest, is not essentially different from that
which other aesthetes might derive from the
paintings of Rembrandt, from the sculpture of
Rodin, from the music of Beethoven, or from
the drama of Shakespeare.

Closely allied with the pursuit of the beautiful is the desire for communion with nature.
For many people there is a significance, as vital
and as satisfying as that which any communicant
ever derives from feeling his spirit one with
God, in feeling themselves to be one with nature,
in cutting all bonds of habit and drifting into
the timeless continuity of the primeval.

The primeval environment often proves exceptionally inspiring and exceptionally favorable to contemplation. Many of our greatest

American thinkers, men of the caliber of Thomas Jefferson, Henry Thoreau, Mark Twain, William James, and John Muir, have found the forest an effective stimulus to original thought. John C. Merriam has well remarked: "The man or woman who goes to Grand Canyon or Yellowstone or Yosemite for a week's vacation is cut off, for the moment, from the business of making a living. Nearly all of the small cares that make up a good part of our ordinary living are also left behind. Under such conditions the mental and spiritual processes turn not only toward appreciation of nature, but toward enjoyment of intellectual life. Frequently, at such times, consideration is given to the greatest of our problems. The most intimate and deepest spiritual experiences may develop and great decisions be made." [12]

Some visitors to the forest are primarily interested in its scientific aspects. They want to study the forest, to learn the fundamental reasons for its development, to appreciate the causes of the functioning of its myriad component parts. To them the forest is a laboratory, unbounded by the conventional four walls, floor, and ceiling of the usual research center, but fully as significant in the development of a knowledge of the laws of nature.

To certain vigorous people there is an intense satisfaction in doing for themselves all the tasks essential for the maintenance of existence. This is never possible under the dependence on others which is inescapably imposed by modern civilization. It is only in the wilderness or on the sea that man can still find an environment where he must be competent to perform all the tasks of life or perish.

The tradition of the pioneers is inculcated in most American children even before they start their schooling. All through their formative years they read about the glorious adventures of the American frontiersmen, and often they relive in games and imagination the stirring pioneer days. Many of them grow up to long for some real adventure and become sick and tired of getting their thrills in such vicarious forms as the lurid movie, the cheap novel, or the travelogue. A depressingly large number of the more energetic of these malcontents try to appease their unfilled yearning in the pursuit of crime and racketeering. Others long for a declaration of war in the hope that in battle they may capture some of the rightful thrills of life. This same psychological urge lures many people to the forest. There, in less antisocial ways than crime and fighting, in the thrill of landing a three pound trout from some mountain torrent, in the competence required to outwit a wary mule deer, in the exhilaration of climbing to some unscaled pinnacle, they add a genuine dash of adventure to their existences.

Finally, there are those whose chief purpose in visiting the forests is simply an escape from civilization. These people want to rest from the endless chain of mechanization and artificiality which bounds their lives. In the forest they temporarily abandon a routine to which they cannot become wholly reconciled, and return to that nature in which hundreds of generations of their ancestors were reared.

As the one environment for realizing these varied purposes, the forest exhibits numerous unique qualities. Of the myriad manifestations of beauty, only natural phenomena like the forest are detached from temporal relationships. All the beauties in the creation or alteration of which man has played even the slightest role are firmly anchored in the historic stream. They

are temples of Egypt, oratory of Rome, painting
of the Renaissance, or music of the Classicists.
But in the wild places nothing is moored more
closely than the geologic ages. The silent wan-
derer crawling up the rocky shore of the turbu-
lent river could be a savage from some prehistoric
epoch or a fugitive from twentieth century
mechanization.

The sheer stupendousness of the wilderness
gives it a quality of intangibility which is unknown in ordinary manifestations of ocular
beauty. These are always very definite two- or
three-dimensional objects which can be physi-
cally grasped and circumscribed in a few moments. But "the beauty that shimmers in the
yellow afternoons of October, who ever could
clutch it." Any one who has looked across a
ghostly valley at midnight, when moonlight
makes a formless silver unity out of the drifting
fog, knows how impossible it often is in nature
to distinguish mass from hallucination. Any one
who has stood upon a lofty summit and gazed
over an inchoate tangle of deep canyons and
cragged mountains, of sunlit lakelets and black
expanses of forest, has become aware of a certain
giddy sensation that there are no distances, no

measures, simply unrelated matter rising and falling without any analogy to the banal geometry of breadth, thickness, and height. A fourth dimension of immensity is added which makes the location of some dim elevation outlined against the sunset as incommensurable to the figures of the topographer as life itself is to the quantitative table of elements which the analytic chemist proclaims to constitute vitality.

A third peculiarity about the forest is that it exhibits a dynamic beauty. A Beethoven symphony or a poem of Shelley, a landscape by Corot or a Gothic cathedral, once it is finished becomes virtually static. But the wilderness is in constant flux. A seed germinates, and a stunted seedling battles for decades against the dense shade of the virgin forest. Then some ancient tree blows down and the long-suppressed plant suddenly enters into the full vigor of delayed youth, grows rapidly from sapling to maturity, declines into the conky senility of many centuries, dropping millions of seeds to start a new forest upon the rotting débris of its own ancestors, and eventually topples over to admit the sunlight which ripens another woodland generation.

Another singular aspect of the forest is that it gratifies every one of the senses. There is unanimity in venerating the sights and sounds of the forest. But what are generally esteemed to be the minor senses should not be slighted. No one who has ever strolled in springtime through seas of blooming violets, or lain at night on boughs of fresh balsam, or walked across damp meadows in early morning can omit odor from the joys of the primeval environment. No one who has felt the stiff wind of mountaintops or the softness of untrodden sphagnum will forget the exhilaration experienced through touch. "Nothing ever tastes as good as when it's cooked in the woods" is a trite tribute to another sense. Even equilibrium causes a blithe exultation during many a river crossing on tenuous foot log and many a perilous conquest of precipice.

Finally, it is well to reflect that the forest furnishes perhaps the best opportunity for pure esthetic enjoyment. This requires that beauty be observed as a unity, and that for the brief duration of any pure esthetic experience nothing remain in a person's consciousness except the feeling of this beauty. In the primeval forest, with its entire freedom from the manifestations of human will, that perfect objectivity which is essential for pure esthetic rapture can probably be achieved more readily than among any other forms of beauty.[13]

CHAPTER V

Forests and Human Happiness

CLEARLY the forests are essential to national welfare. As sources of greatly needed raw material they play a vital part in raising the physical standard of American life. As conservers of soil and water they are absolutely necessary if we are not willing to have our country become as denuded and flood-swept as the Chinese hillsides and valleys. As environment for the highest type of recreational and esthetic enjoyment, they are essential to the happiness of millions of human beings. Economic, physical, and social considerations all demand that we maintain a bountiful forest resource.

Since, therefore, we require forests, and since, furthermore, our present forest policies are resulting in a deterioration of our woodland wealth which is already bringing disastrous consequences, it is patent that we must evolve some new method of handling our forests. Obviously we cannot just say, "Let there be forests", and then find that our command is fulfilled. The forests are delicately balanced biological organizations, the control of which is greatly complicated by the necessity for detailed scientific knowledge. It is also complicated by the many economic and social difficulties which have contributed to the mismanagement that has characterized the handling of our forests ever since the first settlement of the white race on the North American continent.

On the other hand, we must not be deterred from finding a solution through fear of the complexities which inevitably are involved in any change. The working out of a forest program is impossible unless we scorn long established impediments and risk the perils of the unprecedented and imaginative. We must shun conservatism's facile defense of damning any fundamental changes with a glib use of the terms "unrealistic", "impractical", and "utopian". Sensible people everywhere are coming to see that in dealing with economic and social matters we must drop axiomatic notions of essentiality and reason with unprecedented logic on what we are after and how best it may be attained.

What we are after is human happiness. We see that the forests play an important part in promoting human happiness. Our problem then is to find how we may manage our forests so as to realize their highest potentialities for the well-being of mankind.

This problem of management involves more than raising the maximum amount of timber. In striving for that development of the forest which will yield the greatest human happiness, we are obviously after more things than trees. It would be possible to imagine some hypothetical monarch who made a fetish of his forests and superstitiously believed that the primary objective of his regime was to make his timber lands yield as close to the maximum production as the combined labor of his subjects would make possible. In this effort he might make them labor 15 hours a day and go without most of the pleasures of life simply because he felt that every bit of productive capacity should be utilized for the benefit of the forests. Obviously such single-minded attention to the arboreal welfare of the country would not further but decrease the happiness of his subjects. In a less exaggerated way, we must recollect that it is possible for a forest program to lay too much emphasis on the forest itself and to consider insufficiently the related social values.

Therefore, in planning the management of our forests we should consider the effect of our policy on the following major elements in the social scheme:

1. The forest.
2. The land.
3. The worker.
4. The consumer.
5. The community.

There are today in the United States approximately 670 million acres of potential forest land, including merchantable timber, brush land, and abandoned farms. This amounts to more than one-third of the entire area of the country, and is greater in extent than Norway, Sweden, Germany, France, Spain, and Italy combined. We must first of all consider the management of this great domain from the standpoint of finding some method of control which will yield us the forests which we require.

We must also examine the relationship which this management bears to the general problem of land use. Our land is limited. It is so scarce that we cannot afford to allow any of it to go to waste, for the frontier period of land squandering has long departed. Therefore, we must decide what is the most valuable use for every portion of our territory. Perhaps 670 million acres is too large a proportion of a national land area of 1,903 million acres to devote to forests. Perhaps it is not nearly enough. At any rate, it is essential that whatever the scheme of forest organization which we adopt, the possibilities of land planning shall be an integral part of it.

"In 1929, forest work, including logging and wood manufacturing, afforded full time employment or its equivalent to 1,500,999 people. This does not include the merchandising of forest products and services under forest products industries. Neither does it include industries dependent upon forest water, forest ranges, forest wild life, recreation, and minor products and services." [14] With this immense number of people engaged in work dependent on the forest, it is obvious that any successful forest program must consider the welfare of these workers.

The consumer, too, has a major stake in the forest. Whether he uses the wood which it grows, or the water which it regulates, or the beauty which it manifests, he has a tremendous interest which must be safeguarded. It is possible to think of a scheme of forest management which might so lock up the forests that he would receive little of their potential benefits. For instance, the government could theoretically prohibit the shipment of all wood products in interstate commerce in order to give badly depleted forests a chance to build up growing stock, or the government might preclude the use of the forests for recreation in order to decrease the fire hazard. Granted that these steps were desirable from the standpoint of the finest development of the forest, they would undoubtedly still be nefarious from the standpoint of the consumer. This standpoint, therefore, must necessarily be considered in working out a forest policy.

The course of forest devastation in America has made constantly more clear the intricate economic ramifications of the forest. Thousands of communities throughout the United States have grown up as corollaries to the harvesting of timber. Their continuance is absolutely dependent upon the continuation of the forests. The vast number of abandoned towns that are found in cut-over forest regions furnish positive proof of this essential interrelation between growing forests and living communities. Even where towns have not been established chiefly as sources of labor for the woods and mills, they have often been ruined by the destruction of the woods. It has been the experience in countless cases that after the lumbermen have removed everything of value from the forest they have ceased paying taxes and abandoned the lands. This has meant that in order to continue the usual functions of local government, such as the maintenance of schools, highways, and the offices of administration, it has been necessary to raise the tax rate on those sources of taxation which remained. These generally have been the farms of the region. When the taxes have been raised on farm lands the result invariably has been that those farmers who have barely been able to make a living have been driven out of business. Thus the tax base has been further constricted, and this has involved another raise of tax rate on those remaining. As a result another group of farmers, who were barely able to make a go of it at the old tax level, have been forced to abandon their farms. The tax base has, therefore, still further contracted, the rate has consequently raised, and the same story has been repeated again. By this process of pyramiding it has actually happened that entire towns and almost entire counties have been abandoned because the chief taxpayers, the timber owners, have withdrawn their contribution to local government.

With this clear recognition of the vital elements in the problem, we will consider first of all what form of ownership and control would best advance the welfare of each. There are three principal forms which ownership and control may take. One is private ownership, either with or without public subsidy. The second is private ownership with public regulation. The third is public ownership. In the following chapters these forms of control will be analyzed for their effect on the forest, on efficient land use, on the workers, on the consumers, and on the dependent communities. Once the most desirable form of ownership has been determined, it will be far easier to discuss the other steps which will be essential before we can make adequate use of our forest lands.

CHAPTER VI

Private Ownership

ALL through the nineteenth century and the first decade of the twentieth century the timber owners had taken a public-be-damned attitude toward the occasional proclamations of far-sighted idealists that private management of the forests was bringing havoc to public welfare. They held the whip-hand, controlled the legislative bodies in most of the important lumber states, and were perfectly willing to let a few fanatics howl their heads off as long as this did not affect their profits. If an occasional altruistic lumberman did want to manage his lands by forestry methods, he could not compete with the cheap production methods which were possible under the policy of devastation adopted by the majority.

However, after about 1910 the conservation movement really assumed tremendous power in this country, due largely to the enthusiastic leadership of Theodore Roosevelt and Gifford Pinchot. Where formerly there had been only a few lone voices shouting in the wilderness, there now boomed forth a lusty chorus of millions of citizens. The lumbermen could no longer ignore the demands for forest conservation, so they brightly changed their viewpoint.

"We really have always wanted to practice forestry," they now sobbed contritely, "but it is simply impossible under present economic conditions. If you change these conditions, if you make it profitable by tax relief and by financial aid and by helping to improve our markets, we will eagerly practice forestry. We do not want to fight the foresters; we want to cooperate with them."

There were a few foresters who ridiculed this

tone of righteousness adopted by the lumbermen. F. E. Olmsted wrote in 1919:

"For many years the forester has endeavored
to persuade the lumberman that measures for
keeping his lands productive were worth adopt-
ing. Persuasion has utterly failed, resulting in
little more than mild amusement on the lumberman's part. The lumberman, to be sure, has
always expressed a keen desire to 'cooperate'
with the forester in all ways possible; but the
difficulty there has been that in practice the
lumberman's understanding of cooperation has
been to accept everything to his immediate lik-
ing and to yield nothing in return. Precisely at
the point where the forester's suggestions imply
a present restraint on the lumberman's part in
order to insure the future welfare of both the
public and the lumberman, the forester becomes
a theorist, an idealist, and what Chambers of
Commerce are pleased to term an 'insidious influence'. The humor of the situation is that the
lumber industry has fallen to its present level
because of a total lack of theory and ideals. The
time for persuasion has passed." [16]

However, the majority of foresters were op-
posed to Olmsted's views. Filled as their minds

had been from childhood with anti-public ownership propaganda and with notions of the all-conquering potentialities of private initiative, they decided to accept the lumbermen's challenge and aid them through subsidy. In spite of the vigorous opposition of a minority faction headed by Gifford Pinchot, the Society of American Foresters, professional forestry organization, came out officially for the principle of *private ownership with public subsidy*. This principle was given Congressional sanction by the Clarke-McNary Act of 1924, which among other things authorized the appropriation of federal money for the protection of private lands from fire, for the discovery of a way to relieve private owners from some of their burden of taxation, and for the providing of free planting stock and free advice to farmers.

The dominant forestry philosophy of the past ten or twelve years has thus involved the assumption that if the government patted the private owner on the back, and asked him politely to practice forestry, and helped him out with a substantial contribution of money for fire protection, research, and advice in regard to the management of his lands, the private owner would jubilantly and enthusiastically manage his woodlands to the perfect satisfaction of the forestry profession and to the adequate development of the growing stock. "The pressure for better silviculture," wrote one forester serenely, "will come from the changed financial point of view of the timber owner once he understands the possibilities. Surely professional foresters will not be slow to respond to such a demand." [17]

Nevertheless, in spite of all sorts of additional testimonials and the active backing of William B. Greeley, then chief of the United States Forest Service, this splendid theory did not work. There was one serious fallacy which doomed it from the start. Lumbermen do not to any appreciable extent invest their extra profits in forestry. In fact, the more profitable it becomes for them to log, the more heavily they usually cut, and the fewer trees they leave on the ground after their operations are completed. As a result, in many cases where the markets were most profitable there was practically no provision made for a new stand, whereas in the old days, when the market had been poor and economic conditions wretched, a great many small or otherwise unmerchantable trees had been left growing and thus formed a nucleus or a seed source for a new forest. It is true that government cooperation in fire protection did decrease the percentage of lands which were burnt in certain regions, but aside from this one effect, public subsidy and improved economic conditions were virtually valueless in improving the forestry practice on private lands.

The open-minded reader will demand some substantial backing for such a sweeping statement. The United States Forest Service, writing a conservative, scientific report, based on the most comprehensive data ever gathered in the history of American forestry, comes to this conclusion: [15]

"Of the 83 million acres of devastated or poorly stocked forest land, 74 million, or nine-tenths, is privately owned and an appreciable part of the remainder reached this condition before coming into public ownership or as a direct result of private operations and ownership.

"Of the 850,000 acres devastated each year about 95 percent are in private ownership.

"At least 36 million acres of forest are being deteriorated annually, primarily as a result of poor silvicultural practice and unsatisfactory fire protection.

"The extent to which private ownership is responsible is shown by the fact . . . that fully 95 percent of the private cutting is probably made without any conscious regard to the future productivity of the forest, and that nearly all of the cutting on publicly owned forests is designed to perpetuate the forest.

"It is shown further by the fact that nearly all of the 191 million acres of forest land needing but not given protection is in private ownership and that at least 41 million acres, or about 98 percent, of the area burned annually during the last few years, is privately owned.

"Private ownership has held four-fifths of our commercial forest land with from 90 percent or even more of the total potential timber growing capacity. It has held the agricultural land which is being abandoned. It has also held two-fifths of the noncommercial forest land. Practically all of the major forest problems of today have grown out of this ownership. As measured by expenditures only about 10 percent of the constructive effort in American forestry is being made by it. Nearly half of this effort is so remote as to have little or no influence on the forest itself."

This Forest Service report brings out many other matters bearing on the failure of private initiative, even when aided by public subsidy, to preserve the forest values. It finds that the percentage of private lands burned is 11 times as great as the percentage of government lands.[14] It finds, furthermore, that only 0.85 percent of the private forests are managed in a manner which will assure a continual crop of timber,[15] whereas virtually 100 percent of the government lands are managed on such a sustained yield basis.[14] The critical problems of watershed protection are chiefly on private lands.[14] While many people use private lands for recreation when public areas are not accessible, the finest recreational opportunities can not possibly be preserved on private lands.[10]

Since the private owner is thus responsible for almost every serious forest problem, it is natural to inquire what the difference has been between the actual practices of public and private operations. Fundamentally there have been two. The private owner has not been willing to spend money for fire protection while the government has. The private owner has not been willing to cut trees and dispose of brush in a way which will insure a new forest while the government has. These categorical statements need amplification, so I will explain specifically how they apply in two different regions.

In the western white pine forests of northern Idaho the private operator has habitually harvested every merchantable tree. This has meant that he has removed practically all the white pines and cedars down to about eight or ten inches. He has taken the better hemlocks, some of the larches, and most of the white firs. He has left standing a large number of old, defective hemlocks which by their dense shade have precluded any young growth except hemlock from coming back. Actually this has in most cases made little difference because the lumberman has generally not disposed of his brush. This has constituted a terrible fire trap, and fire has usually not been slow to take advantage of the favorable environment. In some cases the owner has disposed of his brush by setting fire to his whole cut-over area and letting it burn broadcast. This kills the young trees which have not been logged, often making a worse fire trap than before. It also precludes any possibility of a new stand through the elimination of both advanced growth and seed source. A brush disposal law has corrected these evils more in theory than in actuality.

The Forest Service practice in the same type of stand has been to leave several large white pine seed trees on every acre. This has generally assured plentiful reproduction of this valuable species. The Forest Service has also preserved the thrifty young hemlocks, cedars, and white firs to put on additional growth, and has only removed the larger trees of these species. The old, rotten hemlocks which compete so severely with the younger trees for light and moisture have generally been either felled or girdled. The brush has been removed by placing it in scattered piles, away from any trees which are left standing, and then burning it. Thereafter the cut-over area has been guarded as carefully as possible from fire. The result in the great majority of cases has been an excellent new forest to take the place of the old.

The huge Douglas fir forests of the Pacific Northwest have almost always been nearly clean-cut by private operators, and those trees which

did escape the axe have generally been knocked
down by the high-power logging machinery.
Furthermore, the highly inflammable débris
left on the ground after such operations has generally burned up after a short time, thus killing
the reproduction which may have started from
the seeds shed by the old stand before it was
logged. The result in about 60 percent of the
cut-over forests has been devastation.[18]

The Forest Service practice in this same sort
of stand has been a gradually developing art.
The present method of cutting is to log only
those groups of trees which are capable of yield-
ing· a definite profit. This means that many
patches of second growth, low-grade old growth,
species not at present merchantable, and inaccessible timber remain as seed sources. When
none of these are available the Forest Service
leaves strips of uncut timber which it will not
log until the adjacent cut-over land shows plentiful reproduction, or it may leave scattered seed
trees standing on the logged area. Slash in clean-
cut areas is generally broadcast burned, but this
is done on days when the duff is so moist that
the minimum damage will be done to the soil.
Snags are often felled, at a cost of three or four

dollars an acre, in order to reduce the fire hazard. Thereafter the land is protected from fire, if not as well as it should be, at least far better than that privately held.

In addition to these specific failures of private as contrasted with government operation, there is the general consideration that stands of merchantable timber cannot normally be raised in less than 40 years, and on the average in not less than 80 years. Therefore, a continuing forestry policy is essential. Private ownership totally fails to provide this continuing policy. A private owner of necessity must consider his immediate profits. If he is a good business man he cannot possibly afford to sink money today in something which will not bear dividends for perhaps 80 years to come. If he has any sense he will realize that 80 years from now the entire economic structure of the country will probably be changed, and that in any case intermediate carrying charges for interest, taxes, and protection costs will eat up all potential profit. So the lumberman invariably calculates that a bird in the hand today is worth a dozen in the bush 80 years from now, and he proceeds to drain the forests of everything that he can possibly market.

Recent studies have shown, it is true, that the leaving of small trees is not only good forestry practice, but good business practice in practically every important timber region of the country. Trees under 12 inches in diameter can usually be logged only at a financial loss, not so much because their quality is poor, as because the amount of work required to produce each board foot of lumber when small timber is being handled is immensely greater than that when large timber is being logged. Consequently, it would seem logical that at least this form of forestry practice, the leaving of small timber, should be adopted by the lumbermen. Actually it has only been carried out to a minor extent, and in many of the places where small trees have been left at the time of logging, the failure of the private owner to provide fire protection has nullified this saving.

So far we have only considered the effects of private ownership on the forests. This has been shown to have been disastrous. The effects of private ownership on the other elements which must be considered in framing an adequate forest policy have been almost as unsatisfactory. It has already been pointed out that careful land planning is an essential feature of any proper forest policy. It is obvious that adequate land planning is impossible under either laissez-faire or subsidized private ownership. For the private owner, except in rare cases of altruism, does not consider what is best for the nation as a whole, but instead takes into account only what is most conducive to his immediate profit. National planning which must be carried out through the voluntary cooperation of private owners whose main objective is personal profit has never yet succeeded anywhere.

The worker in the lumber industry has in many respects been forced to undergo the worst conditions in any industry in this country. In the recent lumber code hearings in Washington operators were seriously defending a five-cent-an-hour wage, and the rate finally proposed was almost the lowest found in any code. The death rate in woods work has been the highest in any major industry, and in mill work it has ranked second. The following table indicates the number of lives lost per million hours of work in a number of the most important industries studied by the Department of Labor: [19]

Industry	No. of Deaths per million hours of work
Logging	1.01
Lumber sawmills	.53
Fertilizers	.49
Shipbuilding	.46
Petroleum refining	.39
Brick	.32
Chemicals	.29
Flour	.25
Agriculture	.21
Iron and steel	.20
Slaughtering and meat packing	.16
Foundry and machine shop	.16
Glass	.15
Automobile	.08
Cotton goods	.04
Boots and shoes	.02

The migratory life of the lumberjack has long been notorious. While this may be partly a result of his own temperament, it is chiefly due to the fact that a person who devotes his life to woods work has almost no chance for a stable existence. The policy of cut out and get out which has been almost universally pursued in the lumber industry has meant that woods operations, instead of focusing around some community, have migrated all over the country, and consequently men who want work have had to migrate with the industry. If the possibility of establishing a permanent home is a source of happiness for the average man, then the unstable conditions of the lumber industry have certainly been injurious to the worker.

"At least," some people insist, "the consumer has profited by private ownership of forests. Devastation we admit, poor land planning we admit, terrible labor conditions we admit, but at least you must grant that the American public has gotten its wood cheaply." It is true that in the past the consumer in the United States has benefited by cheap lumber in comparison to what it has cost in most of the other countries of the world. However, this cheap lumber has not been the result of private ownership, but rather a result of an unusually bountiful natural resource, and a great overproduction. The former is, of course, something for which private ownership deserves no credit, unless under the theory that nature smiles on the righteous. The latter will result in a terrible expense to the consumers of future generations who not only will have to pay the high prices of the inevitable timber scarcity which is coming, but will also have to pay the cost of rehabilitating the millions of devastated acres left in the wake of the cheap logging methods of private industry.

Even without considering the future, the consumer has not enjoyed as cheap lumber as one might suppose he would get because the private timber owner has avoided practically all forestry expenditures. Most of this saving has been swallowed by the wretched organization and management of almost all lumber companies. They have habitually bought up too much land, have been greatly overcapitalized, have as a rule been directed by bull-headed gentlemen with an appalling ignorance of modern business technique, have consistently failed to take advantage of such economies as leaving small trees to grow, and until recent years have suffered from an absence of standardization in production. The cost of this mismanagement has obviously been transferred to the consumer, who has had to pay the necessary interest on the terrific carrying charges of land-poor companies.

The prevalent theory among both timber operators and foresters has been that the greater the demand for lumber the better off our forests would be. Large demand would stimulate price, stimulated price would increase the lumber-

man's profits, increased profits would be invested
in forestry measures, and consequently the
American forests would become superbly thrifty.
It has already been shown that the theoretical
relationship between private profit and private
forestry has not been borne out by actualities.
Furthermore, prior to the depression the national timber cut was twice the timber growth,
and the saw-timber cut five times the saw-timber
growth. Consequently, it seems imperative for
us to reduce our rate of cutting as much as we
possibly can, so that growth may be given a
chance to catch up with the excess depletion.
Therefore, every substitute which can be found
for such major sources of drain as lumber, ties,
and the better grades of pulpwood is in reality
a boon to forestry. Of course, if a person has
money invested in timber, he will not feel the
validity of this argument. This is one more reason why private ownership is especially un-
adapted to meeting public needs. Private owners today want the maximum timber markets
possible, and public welfare demands a curtailment in markets to give the badly depleted forests a chance to recuperate. The depression,
which has virtually wrecked the lumber indus-

try through reducing the demand for lumber from about 36 billion to 9 billion board feet in a three-year period, has been the greatest godsend the American forests have known since lumber production became a major industry.

Almost all public forests are logged on a sustained yield basis which has for its objective the cutting of only as much timber as can be replaced by growth. When this fundamental forestry principle is observed, the volume of lumber in the forests keeps constant, and operations, together with the communities dependent upon them, can go on perpetually. This principle has been almost universally neglected by the private owners, who have consistently cut their timber on the basis of removing everything as quickly as possible and then pulling out of the region. Sustained yield forestry, as practiced by the federal government, is an essential for community welfare, and private owners have had an almost perfect record of failing to realize this goal.

So we see that, from whatever viewpoint we may examine it, private forestry, even when aided by government subsidy, fails to safeguard public welfare. Many private owners, admitting their failure, claim that it is all due to the inequitable system of taxation by which they must pay an annual tax on a crop from which they will receive a return perhaps once in 50 or 100 years. They contend that, if they only had to pay a yield tax when their crop was harvested, then they really would practice forestry. Actually the cost of taxes amounts to considerably less than the charges for protection and interest on investment, even with the present unjust laws. Furthermore, there is no reason to suppose that the lumberman would be more inclined to invest his tax savings in forestry than he has in the past been inclined to invest his increased profits when lumber prices were high.

The only way that private forestry could be a success would be for the government to pay practically all the expense of starting, developing, and protecting the forests, leaving to the owners only the harvesting of the profit. Such a scheme is obviously preposterous, and there is no social justification for the government to use the resources of all the people simply to make possible the continuance of private ownership. Far better it is for the government itself to take over private woodlands and to manage them in that competent way which it has, during nearly 30 years of practical experience, more than amply demonstrated.

CHAPTER VII

Public Regulation

PUBLIC regulation of private forest management commenced in Europe as early as the sixteenth century.[20] It was seriously urged in this country more than fifty years ago by as powerful a person as Carl Schurz, and has been urged repeatedly since then by liberal leaders in all walks of life. In the first two decades of the present century Theodore Roosevelt and Gifford Pinchot stirred public imagination with its possibilities. In 1921, under the urging of

Pinchot, Senator Capper introduced a bill call-
ing for government regulation of private logging
by means of the government taxing power. This
bill was defeated, and the Clarke-McNary Act
supporting government subsidy was passed instead. Nevertheless, a minority of the forestry
profession continued from time to time to urge
the necessity of federal regulation. Several states
passed laws regulating such matters as brush
burning and fire control, but only a few of them
were even of mild help to the forest.[20]

The chief reason for the failure of American
forest regulation has been the bitter opposition
of the lumber companies and their friends. This
opposition has been aided by the fact that the
states with the greatest timber resources were
the very ones in which the legislatures were
largely controlled by lumber interests. These
lumber interests were careful to see that no law
for safeguarding public welfare might find a
place on the statute books if it was detrimental
to the private profits of the lumbermen. As a
result, we find not a single state in the entire
Union with effective laws for the protection of
the forests.

We cannot, therefore, look to American experience for any evidence of what forest regulation might do if adequate laws were passed and enforced. We must turn instead to European experience and make such interpretations as seem appropriate in applying this experience to America. Foreign regulation has of course varied from country to country. In a few nations, of which Switzerland is the outstanding example, public regulation has been eminently successful. It must be recalled, however, that in Switzerland "there are almost no large private commercial forests and practically everyone realizes the necessity of maintaining a forest cover for protective purposes. In France and Germany, the various proposals for new and more restrictive laws indicate that existing legislation is believed to be inadequate." [20] In many of the other countries regulation has proven largely a failure. Sparhawk, in his able study, has summarized the two chief phases of the European regulation in the following words: [20]

"1. Cutting in classified protection forests * may be done only with permission of the com-

* Protection forests are forests which must be preserved for the protection of watersheds.

petent authorities. . . . Reforestation of cut-over areas is compulsory."

"2. The object of control of other than pro-tection forests is generally not to compel owners to produce any particular kind or quantity of material, but to insure that the land will be kept in a productive condition. Sustained yield man-agement is usually not required, and control over methods of management and utilization is reduced to a minimum."

The effect of these two policies of regulation has been in general to prevent actual devasta-tion, to control floods and erosion far more satis-factorily than they have been controlled in the United States, and to safeguard many of the possibilities for recreation. However, they have generally not been sufficient to keep the forests operating on a sustained yield basis. Most of the European countries are cutting more timber than they are growing, and their forests as a whole are marked by steady deterioration, albeit this deterioration is not nearly as critical as that in the United States. The conclusion, therefore, seems justified that in general regulation of the forests in Europe has brought better forestry results than in the United States, but that it has not brought entirely satisfactory forest practice. Indeed, it has only been in those countries, such as Switzerland, Norway, Sweden, and Finland, where the owners favored government regulation of their own volition, that this phase of forest control may be considered a success. In America, where lumber companies have been almost universally hostile to any form of regulation, far more hostile even than in the most backward European countries, it seems nearly certain that regulation would fail to provide adequate safeguards.

American regulation of public utilities is very suggestive in this connection. Mosher and Crawford, although their general attitude is in favor of public regulation, admit nevertheless that to date it has been a failure in the United States. They say:[21]

"The year 1932 marks the closing of the first quarter-century of systematic utility control. It would appear that during this space of time the experimental period would have passed; that the boundaries of public control, as against the freedom of initiative and discretion inherent in private operation, would have been definitely set; that the methods of supervision in the public interest would have become standardized and stabilized. Unfortunately, such is not the case. Making due allowance for the exceptions noted in earlier chapters, it has been shown that the commissions have been swamped with routine work, that they have continued the use of stereotyped methods, that they have been understaffed, and, what is perhaps more significant, that neither the commissioners nor their powers have kept pace with the problems of regulation."

The history of utility regulation in America has been largely one of private interests regulating regulation, both through inadequate appropriations for the accounting activities which are essential in determining just rates, and through getting men friendly to the power interests appointed on state and federal Power Commissions, as well as on the courts. In consequence, we find in the face of theoretical regulation that numerous holding companies in a single year have made between 50 and 3,000 percent dividends upon actual investments in stock.[21] We find the Supreme Court holding that a profit of 7.44 percent on the investment "would be confiscatory and in violation of the due process clause of the Fourteenth Amendment."[33] We find the accounting department of the federal Power Commission so curtailed in funds that its estimates of investments on which both rates and recapture values depend are years behind. We find also that private interests have so powerful a control over federal regulatory bodies that they are able to get an order repealed which would have made impossible the crooked ventures of Samuel Insull.[22] Indeed we find pretty generally such private influence on public regulatory bodies that little is done which is seriously deleterious to private investment.

As a result, many experts who do not feel the urgency of preserving private ownership have come to a great sense of futility in all forms of regulation. There has arisen within the past two years a growing belief that the only way to regulate public utilities is by public ownership. We find that an increasing number of local communities are generating their own electrical energy, that Wisconsin and Oregon have passed laws permitting the state to develop its own public utilities, and that the federal Government has already committed itself to the national development of power at Boulder Dam and Muscle Shoals, and has contemplated similar action on the St. Lawrence River and in the Columbia Basin.

Railroad regulation has on the whole been more satisfactory than utility regulation. It has been more successful in establishing fair rates and has improved the standards of service. It has, however, unquestionably been kept from entirely acceptable functioning by powerful financial interests. Furthermore, it must be recalled that it took 18 years, from the passage of the Interstate Commerce Commission Act in 1888 to the Hepburn Act of 1906, for any teeth to be put into railroad regulation.

If public regulation has thus proved a failure among the utilities and has been only partially successful with railroads, where it might be expected to work relatively simply, the situation with respect to forestry would be infinitely worse. In the first place, most of the regulation of public utilities has been practiced by state boards of control. But for the forests, with each of 27 different states containing at least the significant area of 5 million acres, it would be hopeless to expect all of them to pass adequate regulatory laws. If one state with important timber holdings should pass satisfactory laws and the others should not, the socially minded state would have a hard time competing for markets with the deficient commonwealths, because it is usually cheaper to harvest wood products without the bother of forestry technique and fire protection. As a result, all the states would be virtually driven to the standard of the worst one. To avoid such an intolerable situation it is essential to have all important forest regulations applying uniformly throughout the country, and this would mean federal legislation, something far more difficult to attain than state action.

Even if the necessary legislation were passed, there would be grave doubt concerning its constitutionality. Admittedly there is a growing sentiment in favor of many forms of federal regulation, but just how the Supreme Court would act is a matter which cannot be predicted without positive premonition concerning which justices are likely to die or resign within the next few years. In the case of past decisions, the Supreme Court's opinions in the first and second child labor cases are certainly discouraging to the notion of regulation. However, the Supreme Court has also held that it is within the constitutional power of Congress to regulate

through taxation the coloration of oleomarga-
rine, the sale of narcotics, certain actions of corporations, and the issuance of circulating notes
of persons and state banks. If a law for forest regulation were formulated in order to control water
flow in navigable streams there would be a good
chance that its constitutionality might be up-
held; but the uncertainty is a distinct though not
fatal drawback to the feasibility of regulation.

The ease with which private companies could
sabotage government regulation is another serious obstacle. Forestry is such a delicate art that
its practice requires not only abstract rules but
also people who really believe in them to carry
out these rules. A lumber company could quite
conceivably obey every regulation imposed by
the government to the letter of the law, and yet
for lack of proper spirit leave its lands in a deplorable condition. It would be much easier for
the government to practice forestry itself than
to force some unwilling private owner into such
practice. Forcing a lumber company to practice
forestry against its will is about as futile as forcing a schoolboy with no musical sense to take
piano lessons when he wants to play baseball.

Public ownership is the case of the musical boy getting the lessons he craves.

In view of the fact that the lumber industry today is largely insolvent, it would be virtually impossible for the government to enforce regulation. There would be little which could be regulated by fining a company for faulty forestry practice if the company could not pay the fine. For a regulatory law to be effective it must not only have teeth but also something for the teeth to bite into. In most cases the teeth of federal regulation would have nothing to chew on today except a soupy mass of bankrupt timber owners.

On the other hand, if through some miracle prosperity should return and the lumber companies should become wealthy, regulation would be even more difficult. Experience in this country has shown that whenever an industry becomes prosperous and powerful it grows so strong it regulates its regulators.

Nevertheless, with all its inadequacies, public regulation would far more successfully protect the forests, the land, the workers, the consumers, and the dependent communities than would uncontrolled private initiative. There is no valid social advantage which can be attributed to the latter type of ownership over the former. Furthermore, it will be quite a number of years before the major share of American timber lands can possibly be placed under public ownership, due to the time involved in appraising the forests, coming to terms with the owners, and clearing titles. Pending the completion of a program of nationalization, there is real need for some regulatory laws to stop the most flagrant abuses to private forest lands. There is a rare opportunity for the National Recovery Administration to place in the lumber code an agreement by the operators to manage their lands in accordance with the regulations laid down by the United States Forest Service.

Even if the larger timber tracts all pass into public ownership, there will still be an important acreage in farm woodlands which will remain in private possession for a long time to come. Although farm woodlands have not deteriorated as badly as the large scale commercial holdings, they have nevertheless been running down constantly. In order to stop this retrogression, it is essential that some form of regulation be enforced by the public. Such regulation could probably be carried out by the states far more readily than could regulation of industrial forests, because the products of farm woodlots have a more local distribution than do those of the larger timber holdings. State regulation of these small holdings would obviate the constitutional issue. It might also be obviated in many cases under federal administration if the government made farm loans and mortgage adjustments contingent on its right to regulate. Any public control of farm woodlands should be coupled with public instruction and advice.

In summary it may be said that public regulation of fire, logging, and grazing on private lands would leave the forests in better shape than under uncontrolled private initiative. Furthermore, regulation is the only immediate possibility of stopping the deterioration of small woodlots which will probably remain in private ownership for a long time to come. However, it has generally proven insufficient in foreign countries and has thus far been extremely ineffective in the United States with forests as well as with other resources. There is no indication that its efficacy will be materially increased, and this leads logically to a consideration of the remaining possibility of forest control, public ownership.

CHAPTER VIII
Public Ownership

T HE fundamental advantage of public ownership of forests over private ownership is that in
the former social welfare is substituted for private gain as the major objective of management.
Let us examine how this theoretical principle
has been borne out by practical experience. In
the first place, there is a definite record in practically every country in which the experiment
has been tried that forests are better run under
public ownership than under private owner-

ship. Even where regulation has been in force, the general experience has been that this was not sufficient, and the present tendency in most nations is to accentuate the desirability of public ownership.

The forests which are reserved for recreation must in almost all cases be publicly owned, because practically no private individual could afford to maintain such non-revenue-producing forests. Now if these forests are really to be reserved from cutting it is necessary that sufficient timber for commodity use be raised on the other forest lands. Under the existing character of private exploitation on these other forest lands, we are so seriously depleting their growth capacity through devastation and stand deterioration that unless we have a change in policy there will soon be a most severe timber famine. Experience has always shown that if the balance hinges between satisfying material wants and esthetic wants the choice of the people is on the side of the former. Consequently, if we exhaust forests upon which we depend to produce our material needs it is almost certain that there will be such a powerful demand to invade the forests which we have reserved for esthetic purposes that logging cannot possibly be kept out of the recreational stands. In order to preserve our recreational forests we must maintain the productivity of our commercial ones. As has already been shown, there seems no likelihood of this occurring under private ownership; so, purely from the standpoint of maintaining recreational forests, it would be a great benefit if the commodity forests were also held in public ownership.

General public ownership would make possible a much better coordination of fire protection activities. Private forest lands are so intermingled with public ones that at present the government must protect both alike, for if fire ran wild on the individual's lands there would be no way of stopping it when it reached public territory. The inability of the government to curb destructive logging practice on private lands has resulted in the accumulation of slash and snags which make the problem of adequate protection more difficult not only on the private lands on which they stand, but also on any well managed government land in their vicinity. Fire protection plans in order to be effective must be carried out on a uniform basis over large stretches of territory. In order to do this it is highly advantageous for a uniform type of ownership to exist over these large stretches. Obviously, in view of the completely inadequate efforts of private owners toward protection, the only possible unified type of ownership would be that of the public.

In this connection it is perhaps worth noting that although private owners control 80 percent of the commercial forest lands, they only contribute 7 percent of the total funds available for fire protection. If there is to be adequate fire protection on all forest lands, it is estimated that a total of at least $32,000,000 must be expended annually. In view of the fact that private owners only contributed about $1,000,000 to fire protection in 1932, and since furthermore there seems to be little likelihood of any materially greater outlay from them, adequate national fire protection would make it necessary for the government to raise about 97 percent of the funds. If the government is going to spend practically all the money for protection, it certainly should have the benefit of the ownership of the land.

There are at present in the United States some 83 million acres of non-stocked or poorly stocked forest land. In addition some 55 million acres of abandoned agricultural land should be returned to the forests. It cannot be expected that more than half of these 138 million acres will restock naturally within 40 years, and it will therefore require planting for them to become productive within a reasonable period. The total area planted in the United States in 1931 was 153,460 acres.[14] At this rate it would take approximately 458 years to replant the land in need of planting, even if there were no further devastation. As a matter of fact, in contrast to the 153,460 acres rehabilitated in the year of maximum planting, there were 850,000 acres devastated. Practically all of this devastation occurred on privately owned lands, whereas most of the planting either was done directly by the government or was largely subsidized by it. Obviously, then, any adequate planting program will have to be largely financed by the government. If the government is going to pay the costs of rehabilitating devastated and deteriorated lands it certainly should own those lands in order that the benefits of its huge outlay may accrue to the general public instead of to private owners who, if past experience is any criterion, will proceed to devastate again the rehabilitated land.

In view of the fact that the fastest growing stands require at least 40 years to produce sawtimber, that the average is nearer 80 years, while some of the slower growing forests require at least 150 years to mature, it is obvious that forestry is a long-time enterprise. Now if in the process of raising a stand which may require three or four generations, the owners change every decade or two, as ownership commonly does change in business enterprises, it is clear that the plans of the management will also change. This might mean that owner No. 1 would expend ten or twelve dollars an acre protecting his forests from fire, insects, and fungi, and in making cultural improvements, during the twenty years in which he owns the land, on the assumption that when the stand matures his grandson will be well rewarded. However, he may die, and owner No. 2, who buys it from the family, may decide that money invested in forestry will not pay dividends, and therefore discontinue the protection. The stand is then perhaps destroyed by fire, and all the money invested by owner No. 1 will have been thrown away. If owner No. 3 buys the devastated land a few years later he may decide to plant it, but then Owner No. 4 who takes it from him may refuse to spend the small amount of money needed to weed the gray birch which chokes out the planted pines, and so again the investment of owner No. 3 goes utterly to waste. This is not a fanciful or unusual sequence of events. The same type of thing has frequently occurred in America under the instability of private ownership. The only owner who can be counted on to devise and carry out a long-time program of management is the state. For this reason alone it seems absolutely essential to have government ownership if there is to be satisfactory forestry.

One of the greatest curses of the forest industry today is the terrific over-production of all sorts of forest products. The reason for this is the competition between the many small owners to liquidate their holdings as rapidly as possible, that they may avoid the constant payment of taxes and protection costs and may extinguish standing investments. The only ways to eliminate this disastrous competition would be through combination or monopoly. Lumbermen have argued for the repeal of the Sherman Anti-Trust Act in order that they may combine to cut down production and boost prices. The repeal of this Act and the creation of a huge private monopoly, even if coupled with the right of government regulation, would not at all insure adequate forestry measures, would not protect the interests of labor, would not necessarily protect the consumer's welfare, and above all would not guarantee to the community in which the forests are set a stable industry. The only way to gain the benefits of monopoly, and at the same time insure that the monopoly will be managed to the genuine interest of society, is through public ownership.

A government monopoly would end the unmitigated evils of the present high pressure salesmanship which the forest industries find necessary in order to retain their markets in competition with other materials. Obviously private owners want to make a profit off their land. In order to make this profit they require generous markets for their products. Naturally they have tried to develop these markets, regardless of whether other products might not be better suited for the purpose than wood. Also, they try to develop them regardless of whether the national welfare would not be greatly increased if the wood markets were reduced. As a matter of fact, since we are cutting our wood twice as fast as it is growing and our lumber five times as fast as it is being produced, it stands to reason that our national well-being would be stimulated if we greatly curtailed our consumption. We should take advantage of bricks, steel desks, artificial roofing, and corn stalks whenever these provide as acceptable service as the corresponding wood products. In this way we not only give the consumer as much satisfaction, but also give our over-exploited forests a chance to recuperate. This elimination of forced markets is obviously not possible under a competitive private industry. It can only be accomplished under public ownership.

The status of government forest workers, whether they be technicians, clerks, or common laborers, has on the whole been better than among private industries. It is true that until the depression the salary paid to technicians was generally lower and to common labor no higher than that paid by private concerns. However, other essentials of a living standard, such as agreeableness of working conditions, regularity of employment, safety of the employees, and compensation in case of accident have been much higher under the government.

One factor which especially tends to break down the standards of the lumber worker is the irregularity of employment. Production under private ownership is chaotic. Mills are constantly opening up and closing down, forests which had given employment to thousands are suddenly wiped out by fire, and regional devastation necessitates the wholesale migration of the lumber industry. In order to remedy such disorganized working conditions it is important that some stable form of management be set up which can provide a basis for adjusting the number of workers to the actual needs. Such a balanced control is obviously only possible under the stability of public ownership.

Rural stabilization and rural planning are both dependent on the sustained yield management of the forests. Unless our timber growth balances our timber cut there is no way of bringing about a stable community life in those regions in which the forest is an important source of wealth. The only forests in the United States which have ever been managed according to sustained yield principles, with a few almost insignificant exceptions, are the government-owned forests. Therefore public ownership seems an essential to rural stability in forest regions.

It is interesting to note that many of the often quoted disadvantages of public ownership are now exploded. We have always heard that one of the great benefits of private ownership was that it cost the people far less than the inefficiencies of government operation. The United States has for more than 30 years been managing between 100 million and 160 million acres of national forests. All the phases of its administration have been models of efficiency and service, far superior in their way to any similar tasks performed by the lumber companies. It is only necessary to visit a government fire camp and a private fire camp, to see a government planting job and a private planting job, to observe the activities of a government ranger and a private ranger (in the rare cases where one exists), to realize just how much better the government has performed the forest tasks than have the private owners.

There is the further point that the government today provides five-sixths of the amount of money being expended on private lands for protection purposes. In return for this donation it has demanded no guarantee of forestry practice. Yet in spite of the subsidy and the privilege of wrecking their lands, the private owners generally have not been making profits. Indeed, most lumber companies are today on the verge of bankruptcy, if they have not already failed. Their management of the best lands in the country has not been efficient, even from the standpoint of private profit.

Rauschenbush has shown that the relatively well managed power industry would be cheaper to the consumer in the long run if the government went to the huge expense of purchasing outright the entire privately owned developments.[23] It has been almost universal experience that public power operations have furnished far cheaper electricity than have private ones functioning under similar conditions. Mosher and others, comparing the cost of electrical energy to the consumer in six Ontario and eighteen New York cities, found the cost of service to be from one-and-a-half to four times as much in New York.[24] The Canadian cities had their electricity supplied by the publicly owned Ontario Hydro-Electric Commission, while the American consumer had to buy his power from private utility corporations. It is true that the private companies paid higher taxes than did the public commission, but through receiving electricity at cost the Ontario municipalities saved enough money to make their net return almost three times as great as that of the New York state cities. With forests there is a further charge against private ownership, the huge bill which must sooner or later be paid for rehabilitating the land which has been wrecked under the mismanagement of individualism.

In the final analysis, public ownership of forests is absolutely necessary, if for no other reason than that the nation is headed straight for forest bankruptcy under private ownership. Ahern has shown that every important timber producing state except Maine and New York is cutting more lumber than it grows.[25] Devastated acres keep piling up at an appalling rate, while drastic deterioration affects several times the area that devastation ruins. It is estimated that the forest drain is 1.8 times the forest increment, while among stands capable of being harvested for sawtimber the drain is 5.1 times as great as the growth.[14] The end result of such a tendency is inescapable.

At the risk of being banal it is worth pointing out that forests are a natural example of capital and interest. The capital in the case of the forest is the trees, and the interest is the annual growth which they put on. The greater area you have with well distributed trees the more interest you will get, for each tree, if not too crowded, will add a certain amount of growth every year as long as it lives. The highest interest rate in the forest always comes from young stands, which may sometimes add in volume as much as four or five percent in a single year. On the other hand, mature stands grow slowly, and often the interest from their small increment is counterbalanced by the capital loss incurred when large trees die. That is why foresters say that an old forest is an unprofitable one, because it pays no interest, but merely maintains an approximately balanced capital. However, where foresters have often erred in their reasoning has been that a balanced capital without interest is far superior to destruction of all or a large share of capital, and consequent loss not only of the capital value, but also of the contemplated interest.

The result of most of the private forest practice in America has been at best to substitute sparsely stocked immature stands for heavy mature stands in which growth was practically balanced by decay. In these younger stands decay has been eliminated, but the number of trees on which growth could occur has been so limited and the rate of growth had been so reduced through the creation of an unfavorable environment, that the new forest, though gaining in volume all the time, actually has held no hopes of ever approaching the volume of the old stand.

All through the forests of the United States, almost wherever logging has occurred, there has been this same serious depletion of capital. We do not have enough trees in the country to make it possible to grow the amount of wood we consume annually. Year by year this inadequate number of trees is being further diminished by the continued process of cutting more than we grow, and so year by year our capital decreases and with it our hopes of income. We are like the spendthrift son of a multimillionaire who cannot live on the interest of his father's fortune, and proceeds to dig ever more deeply into the capital, thus cutting down year by year the interest he receives, and approaching constantly closer to that insolvency which can be the only culmination of his extravagant policy.

How much land should we have to put under forestry management in order to end this constant depletion and grow the timber which we require? We have in the United States 670 million acres of potential forest land. On 109 million of these acres the environmental conditions are so unfavorable that it would not be possible to grow merchantable timber, while 11 million additional acres of good forest have been set aside for recreation. There remain 495 million acres of forest land and 55 million acres of abandoned submarginal farmland, which are productive enough to raise stands of commercial timber. From this total of 550 million acres there must be deducted 41 million acres which should be set aside for special purposes, chief of which is recreation. This leaves a total of 509 million acres at present available for the growth of forest commodities.

The Forest Service has not had the pessimism to calculate what the result would be if present forest land management were projected into the future. It has, however, made calculations of how much timber we might grow on these 509 million acres by the year 1950 if we adopted various degrees of improved forestry.[1, 26] The minimum method which the Forest Service has considered involves putting forest fire protection on a satisfactory basis, making a large curtailment in the over-cutting in the Eastern regions, and expanding the areas now under intensive and crude forestry to about 10 million and 100 million acres respectively. Under this scheme, assuming that the total drain on the American forests would be cut to about five-eighths of the present figures, it would be possible to grow approximately 10,620,000,000 cubic feet of timber annually. This would be far less than we were consuming before the depression, and would leave no opportunity for recreational expansion.

If we indulged in the maximum forestry scheme investigated by the Forest Service, including the practice of highly intensive forestry on at least 100 million acres and somewhat less exacting forestry on 339 million additional acres, we should be able to grow about 21,450,000,000 cubic feet a year. This would be approximately one-quarter again our present production and would perhaps be considerably more than we should have any need to consume. Nevertheless, as I have already pointed out, the chances are that if we had such an over-abundance of timber we should find many new uses for this readily available product, not to mention the possibility under those circumstances of greatly increasing our lands available for recreation.

We certainly need something approaching this second plan if we are to avoid a catastrophic depletion in our timber wealth. Under private ownership we could not nearly hope to attain the first plan. Years of experience both in this country and in Europe prove conclusively that it is impossible. The government cannot afford any longer to stand idly by while private owners wreck the nation's wealth. Only through immediate, large-scale public acquisition will it still be possible to save the nation's forests.

CHAPTER IX
Method of Acquisition

SINCE public ownership should be the major system of forest control in America, it is first essential to decide whether it is wiser to confiscate the land which needs to be acquired by the government or to purchase it. Whatever a person's philosophical belief may be concerning social rights and property rights, the fact is inescapable that with the country functioning on a capitalistic basis it is out of the question to consider confiscation as a feasible means of acquiring public forests. In the first place there would be general objection to the arbitrary confiscation of the wealth of just one group of owners, and in the second place it is absolutely certain that the Supreme Court would declare such confiscation unconstitutional. If we plan to nationalize our forests while the country as a whole is operating under a capitalistic system, we shall have to pay for them.

Before going into the details of acquisition, it is important to divide the 670 million acres of potential forest lands into three classes. These are:[27]

Abandoned agricultural land	55	million acres
Non-commercial forest land	120	" "
Commercial forest land	495	" "
Total	670	" "

If the government desired to wait for a couple of decades it could undoubtedly acquire virtually all of the abandoned agricultural land for non-payment of taxes. However, in view of the fact that erosion is rapidly deteriorating much of these areas, it would be far more economical for the government to bear the cost of buying them immediately and thus to save the greatly increased expense of rehabilitation which any delay would entail.

The non-commercial forest land embraces 11 million acres of good timber devoted to recreation, and 109 million acres of brush land, piñon and juniper forests at the edge of the southwestern deserts, and scrub growth at high elevations. This involves no serious problem of acquisition. At present some 62 million acres is in public ownership and 58 million acres in private ownership. There is no necessity for the government to purchase any more of the privately owned brush land than is needed to assure adequate protection of those areas which have a major influence on stream-flow or erosion. These areas embrace approximately 30 million acres.[7] The remaining 28 million acres can well remain in private ownership, provided they are placed under satisfactory fire control.

The commercial timber land in the United States may be subdivided into three main classes of ownership. These are: [27]

Farm woodland	127	million acres
Private industrial ownership	269	" "
Public ownership	99	" "
Total	495	" "

The farm woodlands cannot generally be nationalized without also nationalizing the farms. Millions of farmers depend on them not only for their fuel, but principally for the small income obtainable during the winter months by cutting and selling such products as cordwood and fence posts. To take away these woodlands with their opportunity for profit would be tantamount to driving a large fraction of the farmers of the country from their agricultural lands as well. Until the time may be ripe for public ownership of all land, it will be impossible for the government to take over a large proportion of the farm woodland. However, a good many million acres of these lands are already tax-delinquent, while several million additional acres will have to be acquired by the government at all costs to stop erosion. Nevertheless, at least 80 million acres of the present 127 million acres could safely remain in private ownership.

The private industrial forest lands, although they were originally covered by splendid forests, today bear many grades of timber or lack of timber which may be classified in the following way: [27]

Saw-timber (stands merchantable for lumber)	90 million acres		
Cordwood (stands chiefly valuable for fuel, pulpwood, etc.)	62	"	"
Reproduction (fair to satisfactory)	63	"	"
Devastation (or almost so)	54	"	"
Total	269	"	"

It is of special importance that the saw-timber stands of the country be placed under adequate forestry practice before private mismanagement so damages their productivity that the government will be forced to spend huge sums in its restoration. It would be a great deal cheaper for the government to purchase immediately the 90 million acres of saw-timber than to wait for its exploitation and abandonment. Virtually none of this saw-timber area should be left in private ownership, which almost certainly results in severe deterioration if not in outright devastation.

Of the cordwood area, probably one-third is already tax-delinquent or will be within the next few years. The remaining approximately 40 million acres should be purchased by the government. On much of the cordwood area there are enough good trees to make excellent stands if the defective trees and the inferior species, which tend to crowd out the good ones, are removed in the near future. But these lands are steadily deteriorating because of this competition of inferior species and, even more, the constant damage by uncontrolled fires. Again, it would be far cheaper for the government to buy outright at existing stumpage prices than to incur eventually the terrific expense of converting almost worthless areas of weed trees back to productivity.

The logged and burnt lands on which a fair to satisfactory number of new seedlings have germinated will require many decades before they can possibly yield their owners any profit. Consequently, few private owners are willing to give these tracts the protection required for them to bear mature forests. Indeed, one cannot blame an owner who is in the timber business to earn his livelihood for refusing to invest a lot of money in land which at best will only yield a profit to his grandchildren. Because the public alone can be expected to maintain a long enough life to reap the benefits of money invested in such lands, virtually all of the 63 million acres in this category should be taken over by the government. About half of this area has already reverted to the government for tax delinquencies, or will so revert within the next few years.

Finally, the 54 million acres which are either devastated or are almost in that deplorable condition should be taken over by the government immediately whenever they have not already reverted for non-payment of taxes. For these worse than worthless areas absolutely nothing should be paid. The erosion which takes place upon them, their contribution to heightened flood crests, and their painful extinction of scenic values is so pronounced that the legal machinery could readily be established for the government to confiscate these mismanaged lands and straightway undertake the task of rehabilitating them.

It is important to stress that any timber and any land which the government purchases should be bought at the existing stumpage or land prices. The lumber companies will unquestionably want to get something approaching the bloated prices of pre-depression years, but there is absolutely no justification for the government paying more than these lands are worth today. Anything additional would amount to a subsidy to private owners as a reward for their years of devastation, and would be unthinkable except to the lumbermen themselves. In view of the deplorable financial condition of the lumber industry, which calculates a total deficit of 168 million dollars in 1931,[28] it is almost certain that most of the companies, however strongly they might hold out for inflated values, would eventually accept almost any offer which would give them a modest but certain return in place of the mounting deficits they now have to meet.

Figured from a rough estimate of average 1932 prices, the total cost of the different types of land which the government should purchase is approximated in the following table:

Type of land	Million acres to be purchased	Average cost per acre (dollars)	Total cost million dollars
Abandoned farm	30	2.50	75
Non-commercial	30	1.00	30
Farm woodlands	17	2.50	42
Saw-timber	90	8.00	720
Cordwood	40	2.50	100
Fair to satisfactory re-stocking	33	1.50	50
Devastated	0	...	0
Total	240		1,017

In addition to these 240 million acres to be purchased, it is estimated that about 161 million acres will come back to public ownership within the next few years through tax delinquency. This area would include 54 million acres of devastated forest, 30 million acres of fair to satisfactorily reproducing forest, 30 million acres of farm woodland, 25 million acres of abandoned agricultural land, and 22 million acres of commercial cordwood.

The total of approximately $1,017,000,000 which it would eventually cost the government to nationalize the bulk of its forest lands is a large sum. When one contrasts this, however, with the sums appropriated for various other projects it really seems like a modest expenditure, for the government has already spent $165,-000,000 on the Hoover Dam, $500,000,000 on the Panama Canal, and $2,000,000,000 on river and harbor improvements.

The payment for this large scale public purchase of forest lands would not have to be made all at once. Some scheme could be adopted similar to what has already been suggested for the purchase of the New York City subways, and such as Cole [29] recommends for nationalization projects in general. In these schemes the government only pays a share of the purchase price each year, and this payment must come from the proceeds of the property purchased. Specifically, in the case of forests the government would issue notes to timber land owners for the present value of their property as estimated by the government, these notes to be paid off at the rate of 4 percent per year for 25 years. This would mean a payment of about 41 million dollars annually. It would be guaranteed for the first five years out of funds to be provided from the Treasury if necessary. Thereafter, the former owners would have to take a chance on getting their returns from the net proceeds of the commercially operated government forests. If the government could not make a 4 percent profit on these lands it would be due chiefly to the great expense to which it would have to go in blotting out the past ineptitude of private operators. Should this expense make the forests purchased from private owners a liability instead of an asset, then it would not be right for the government to pay a profit to the people who were responsible for the liability. In other words, after the first five years of assured income, an inducement worth offering in order to bring about immediate nationalization of the land, the private owner's income would be contingent upon the government finding itself in possession of property from which income might be reaped.

If the purchase program here recommended were carried out, the following would be the approximate distribution between public and private ownership of forest land:

Type of land	Public Owner-ship Million Acres	Private Owner-ship Million Acres	Total Million Acres
Commercial forest	415	80	495
Non-commercial forest	92	28	120
Abandoned agricultural	55	..	55
Total	562	108	670

Today the 99 million acres of commercial forests in public ownership are divided into 88 million acres of federal land, 10 million acres of state land, and 1 million acres of county or town land. In general the counties are not interested in administering forest lands, except occasionally for local parks. Almost without exception they are in very desperate financial straits, and they cannot be counted upon to pay any part in the huge purchase program proposed. As between the federal government and the states, it may be said in general that when the states are willing to spend the funds necessary to carry out adequate management, then it is satisfactory for them to own the lands. However, federal ownership is generally better than state ownership because it is further removed from politics, because it is not so subject to the control of private lumbermen, because greater federal funds are available, and because under federal ownership national planning is most readily possible.

With the probability of 160 million acres of tax delinquent lands within the next few years, one of the most important steps toward public ownership is the immediate setting up of the machinery for taking over this territory which is coming to the public for nothing. At present this has become a no-man's-land, an unadministered public domain for which nobody feels any responsibility, a vast kingdom which is constantly deteriorating under the influence of fire, erosion, and the competition of weed trees with valuable species. According to the laws at present in effect, the title in tax-delinquent lands reverts to the towns or counties in 29 states and to the state itself in 19. In most cases neither the towns, the counties, nor the states have made any provision for the administration of this new territory. It is imperative that laws be passed at once providing for the automatic transfer of tax-delinquent lands from the control of the smaller political units into state or preferably federal forests. The states or the federal government should thereupon set up funds for the proper management of these abandoned lands. It would be greatly to the advantage of the counties or states to give these unadministered lands to the United States for nothing but a guarantee of management, and thus make them a source of wealth rather than of liability to the local population.

This brings up the question of how local governments can function without the taxes which they have been accustomed to collect from private forest lands. An arrangement could readily be made, similar to what is now adopted with the national forests, whereby 25 percent of the gross income is returned to the counties in which it was made and an additional 10 percent of the government proceeds is invested in road and trail developments. Through these definite returns from well managed federal lands the counties are actually better off financially than they would be with the irregular and constantly diminishing returns from the taxation of private forest owners.

Kneipp [30] has made a thorough study of the benefit to state and county governments of national forest ownership. He shows that if the national forests had been privately owned during the years from 1923 through 1927, at the most favorable period for the successful operation of private forest lands, they would have only yielded through taxation a net gain of $3,037,-436 to the states and counties in which they were located. Under federal management these same forests actually contributed $10,245,394 in direct revenue and aid. In addition they were worth approximately half a million dollars more through the benefits accruing to state, county, and private lands from the fire and insect protection activities on the national forests. Clearly, federal ownership of forests instead of working a financial hardship on local government, greatly increases the state and county revenues.

Even with public ownership of the forests

there is no assurance that their values will be exploited for the benefit of all the people. It has frequently happened that a few private individuals have reaped most the profits from public ownership. The activities of the United States Shipping Board, the leasing of federal power sites for a small fraction of their worth, and the common practice by which municipally generated power is distributed at great profit by private utilities are all examples of public ownership in which a considerable share of the public values has been perverted to pay a profit not to the general run of people but to a few notoriously wealthy ones. In order that this injustice may not also occur with the forests it is desirable that the government should not only own but also operate its timberlands. While the acquisition program could go through without provision for public operation, the public would under such circumstances lose a large part of the benefits which it might reasonably expect to derive.

Before public ownership can be carried beyond the woods and into the mills, it will be necessary to overcome one almost insurmountable obstacle. The value of all the American

lumber and paper mills, even at depression figures, would run well into the billions of dollars,
an amount of money ridiculous to contemplate
when added to the amount that is needed to nationalize the forests. Furthermore, the lumber
mill capacity of the country was approximately
twice as great as the consumption in the last pre-
depression year of 1929,[28] and so the government, if it purchased all the mills, would have to
pay for much useless equipment of an over-in-
flated industry.

Even if all the lumber and paper mills were
taken over by the government, there would still
remain in private ownership a vast chain of industries which depend upon the forests. More
than 50 million board feet of lumber is remanu-
factured annually into each of the following
products: boxes and crates, furniture, cars,
vehicles and vehicle parts, woodenware, caskets
and coffins, refrigerators, agricultural implements, fixtures, ships and boats, handles, matches
and toothpicks, musical instruments, tanks and
silos, and electrical apparatus.[2] In addition to
these products, practically all of the lumber
which is used in building and construction must
pass through the hands of the wholesaler, re-

tailer, and the contractor, and yield each a profit. Paper used in books passes from the mill to the printer to the publisher to the bookstore before it is eventually acquired by the reader. At the present time there would be almost as much reason to nationalize all these activities as to nationalize the mills, and the cost would obviously be prohibitive.

If the government owned and operated the forests of the country it could exert a powerful influence in remedying some of the major inefficiencies in the lumber industry. It could readily eliminate the unmitigated evil of overproduction by refusing to sell to the mill owners any more logs than the market required. The most important reason for overproduction today is the necessity for private owners to liquidate forest holdings before all chance of profit is eaten up by protection costs, taxes, and the huge carrying charges, brought about by overinvestment in mills, equipment, and timber during a period of rising prices. This cause would vanish if the government owned the timber, and production could then be held down to the basis of legitimate use. The government could also eliminate much mill waste through confining its sales of logs to those mills which would make efficient use of raw material.

Similarly, the government could, if it had the will, improve the status of the mill workers by refusing to sell any timber to companies which would not maintain prescribed wages, hours of labor, and working conditions.

In summary, it should be stressed that the immediate necessity is to save the forests of the nation. In saving these forests time is truly the essence of the problem. Under a form of government which forbids confiscation and requires purchase, the rate of forest acquisition cannot afford to be diluted through the distraction of funds which would be necessary for the purchase of the mill properties. After the forest lands which are now in the process of rapid deterioration have been acquired, it will be time enough to consider the desirability of also purchasing the mills.

CHAPTER X

Land, Labor, and Livelihood

PUBLIC ownership solves many of the problems of the forest, but it is not a panacea. There
are some special complications which must be
met even after the forests have been nationalized.

The first of these is the problem of land use.
There are in the United States 1,903 million
acres of land. It is estimated that originally some
820 million acres of this total was covered by
what we would call commercial forests. Today
this has been reduced to 506 million acres. If

present tendencies of devastation continue it is not unreasonable to assume that the commercial area by the year 2000 may have declined to 300 million acres. The question arises, Which of these is a desirable proportion of our land resources to devote to forest growth? Will 300 million acres be sufficient, or do we need all of our present 506 million acres, or should we return to the original 820 million acres? Or are our forests so vitally important that we should plan on putting every acre of the country into timberland?

It is true that the more land we have in forest growth, the cheaper will be our lumber, the better will our streams be regulated and our soil conserved, and the greater will be our recreational opportunities. If these things embraced everything of importance in our social order, we would say immediately: "Put all of our land into forests!" But of course these things do not embrace everything. We need land for crops, for pastures, for houses, for roads, for golf courses, for cemeteries, for garbage dumps, for zoos, and for scores of other uses. If all of our land were devoted to forests, we should be immensely

poorer through eliminating these values from
our development. Our civilization would be laid
on a par with that of some of the bushmen who
never come out to the sunlight.

The question next arises, How much land
could profitably be covered by forests? The logical answer would seem to be that all land which
does not have a definitely better use should be
kept in tree growth, because as already explained,
the more land we have with forests the better off
we are, provided that some other important land
use is not sacrificed. At present the chief land
use competing with the forest is agriculture. A
few generations ago agricultural lands were making steady inroads into the forest, but today the
reverse is true. Already there are 55 million acres
of abandoned farm land in the country, and the
Bureau of Agricultural Economics estimates an
auditional abandonment of between 25 and 30
million acres in the next 20 years.[31] Clearly the
present trend is from agriculture to forests. The
question today is not whether we should steal
land from other uses in order to make forests,
but whether it is better to enjoy the manifold
services of the woods than to drift into the lazy
acceptance of idle land.

Even though we accept the general principle that all land not otherwise in use should be devoted to forests, it is still desirable to undertake a careful land survey of the entire American acreage so that we may determine which areas are best adapted to different uses. Such a survey would reveal certain very rich forest land which could be converted to agricultural use, in spite of general trends. It would also indicate certain abandoned agricultural land on which farming should be resumed, and much agricultural land, not yet abandoned, which should be planted with trees or otherwise treated in order to stop erosion. This work would demand not only expert foresters, but also trained geologists, ecologists, soil scientists, and economists. It would unquestionably require a considerable outlay of money, but by eliminating the present chaos wrought by our almost complete lack of land planning, it would be a small expense.

Under public ownership the government should establish standards of wages, hours of labor, and safety of working conditions far above those now prevalent in private industry. The government is supposed to be operating for the

welfare of all the people in contrast to private
industry, which is operating primarily for the
profit of the few who have money invested in it.
Since the majority of people are wage workers
or their dependents, it follows that the government, in raising their standard of living, is functioning for the benefit of the majority of those
for whom it is supposedly created.

Although this principle is generally accepted
philosophically, in practice the government has
not always been too generous in the payment of
its workers. While they have on the average enjoyed better standards than men in comparable
private jobs, it was a notorious fact before the
depression that technicians were often offered
several times the salaries they were getting from
the government as an inducement to go into
private work. As for common labor, until the
depression the government never paid its woods
workers any more than wage rates standard in
lumber camps, while for a time during the
World War it paid a great deal less. This meant
that even at the height of prosperity $2.60 and
board was the maximum paid for an eight-hour
work day in the woods. On forest fires, the
United States Forest Service has customarily

paid 30 cents an hour, from which $1.20 a day has been deducted for food. Thus for a twelve-hour day of the severest labor imaginable, with many dangers involved, the government fire fighter would generally make only $2.40. The excuse for these low wages was that anything higher would tend to induce incendiarism.

In order to give the workers on the forest a reasonable chance to protect themselves, it is necessary that collective bargaining be recognized in government work as it already is in many of the private industries. Only through collective bargaining can the workers be assured that their interests will receive any consideration. It is presumed that the government will continue to safeguard the forest employees through workmen's compensation and adequate hospitalization. It should also strive to cut down the danger of forest work, which is more perilous than any other major type of work in the country.[18]

The permanent jobs in the organization which will administer the public forests should, of course, be under civil service, regardless of whether these forests are in federal or state control. With all its flaws, civil service seems to be the best method yet devised for keeping up

standards in employment, and for preventing
the government service from degenerating into
a political hodge-podge. The superior quality of
personnel in those branches of the government
under civil service furnishes an eloquent testi-
monial to the desirability of a civil service status
for government employees.

There are many problems of rural livelihood
which could readily be resolved under public
ownership, but which would not necessarily be
met without special planning. In other words,
public ownership would furnish an ideal background for meeting these problems, but by itself
it would not meet them.

In most rural parts of the country the sub-
divisions of government are bankrupt. They
have expended such huge sums of money on
roads, on schools, and on local graft that they
would have difficulty in raising the necessary
funds even in prosperous times. With the depression it has been absolutely out of the question to obtain from the poverty-stricken farmers
the taxes required for the upkeep of local government. Combined with this has been the
abandonment by lumber companies of tens of

millions of acres of land which formerly paid a good share of the taxes. These timber-lands have been so wrecked by private mismanagement that they will not be able to yield an income for generations to come. The possibility, therefore, of meeting the large deficits of local government seems exceptionally remote.

The only solution is complete rural reorganization. Many entire towns and even counties should be abandoned to the forests. In them there is so little possibility of agricultural profit that what small income might be realized would not be enough to pay the expense of maintaining the necessary roads and schools. Rural government has been a terrific burden on the rural taxpayer chiefly because land ownership has been so inefficiently distributed. It costs very little more to maintain a good road which is used by 3,000 people than it costs to maintain one used by 30 people. Yet the drain on the income of the 3,000 people in maintaining such a road would probably be of little moment, while with 30 people it might well prove ruinous. If the agricultural population could be concentrated in those areas best adapted for agriculture instead of being scattered all over the outdoors, the per capita cost of maintaining roads and schools would be immensely decreased. The first necessity in rural reorganization is, therefore, so to distribute population that the maximum of governmental services may be obtained with a minimum of taxes. This could be arranged through offering to give farms in sections where it is worth trying to maintain communities in exchange for farms in regions which, in the judgment of the land survey, should be abandoned to the forest. A step in this direction has been taken in the recent Wisconsin statute which empowers the county boards to "regulate the areas within which agriculture, forestry and recreation may be conducted." [32]

There is quite a rational question whether it would not be better to eliminate all local government and have the state or nation perform the duties now carried out by counties and towns. Because of a tremendous duplication in the offices of government, many people must pay taxes to five or six different agencies.[34] With the general bankruptcy of towns and counties all over the country, and the consequent necessity for the states or the nation to subsidize them if they are to continue functioning, the question naturally arises whether it might not be better for the larger agencies of government to perform the services for which they supply the money. The only serious argument against such a course is the prevalent belief that the more closely related the government is to the people, the more efficiently it is carried out. However thoroughly this notion may be ingrained in the American mind, actuality indicates that towns and counties are by far the most inefficient of the governmental agencies in the United States. With this argument invalid, with the additional advantage of overhead reduction in large scale activities, and with the greater possibilities for nationwide planning, it would seem that the welfare of the rural population would be greatly increased if local government were surrendered to the federal or state authorities.

Most farmers cannot make a living today on the products of their farms alone, and they are seriously dependent on any subsidiary income which they may be able to acquire. Furthermore, the government is undertaking the experiment of establishing some of the excess industrial workers on farms where they will raise crops not for the market but for subsistence. In order that they may purchase those goods and services which can not be produced on the farm it is essential that they be given either some wage work or a dole. It would be of tremendous benefit to both of these classes of farmers, wherever they live in forest regions, to be able to count on a few weeks' work each year in the woods near where they live. Under the planned economy which would be possible with government ownership of the forests, it would be perfectly feasible to distribute the forest work so that each farmer would get his share of the potential wages. At the same time he could continue to live at his neighboring home and not be forced to sacrifice family and stability, values of life which the woods worker must generally give up under the prevalent mismanagement of the forests.

In order to make the forests a constant source of wages to those who live in their vicinity it is essential that they be placed under sustained yield management. This fundamental forestry concept simply means that no more wood may be cut from the forest in a given year than the forest is capable of returning in growth. Only if the woods are managed on such a basis can there be any stability to rural livelihood in forested regions.

CHAPTER XI

Recreational Safeguards[10]

PUBLIC ownership in itself is not enough
to safeguard the interests of the recreationist in
the forest. It is quite conceivable that under
public ownership the forests might be administered too largely from the standpoint of producing raw material, while the esthetic values
might be largely disregarded by those in control.
As a matter of fact, this very situation has arisen
in the management of the national forests, for
their administrators have exhibited an amazing

[171]

hesitation in setting aside any large areas of virgin timber to be reserved from cutting. Even if the recreational values were fully recognized by the government, there would still be many intricate problems of how the recreational areas should be handled.

It has previously been pointed out that those who seek the forests for recreation come with many different purposes. In order to satisfy these varied objectives they take their recreation in a wide variety of ways. Some travel exclusively by auto, some take short strolls back into the woods, some lose themselves for weeks in the primitive conditions of the wilderness, some spend their vacations at luxurious resorts, some care for hunting and fishing, and some demand the thrill of perilous mountains. Obviously, these diverse recreational activities require for their realization several different types of forest. Since each of these types has distinct standards of size, beauty, and administration, and since therefore a separate management plan must be developed for each, it is well to consider them separately.

"Superlative areas" are localities with unique scenic values, so surpassing and stupendous in their beauty as to affect almost everyone who sees them. Examples of such superlative areas are the Grand Canyon, Crater Lake, the Canyon of the Yellowstone, the mountains of Glacier National Park or the Teton Range, the valley of Yosemite, and the Big Trees. If any of these areas should be destroyed or seriously injured there would be no substitutes.

Most American citizens are enthusiastically convinced of the importance of thoroughly safeguarding the inspiring beauty of these areas. Even those least appreciative of recreational values are generally willing to admit that we should reserve our relatively few superlative areas. Consequently it will not be necessary here to justify the retention of such tracts for recreational purposes. Already most of them have been set aside as national parks. A few which remain in private ownership should be acquired by public agencies as soon as possible.

"Primeval areas" (sometimes called natural areas) are tracts of virgin timber in which human activities have never upset the normal processes of nature. They thus preserve the virginal growth conditions which have existed for an inestimable period.

Primeval areas have two major values. First, they are of great scientific significance. They are an absolute necessity for any future studies of the natural distribution of the flora of the world. Further, in trying to plan methods of cutting which will assure the perpetuation of the forest, it is of utmost importance to have various unmodified stands with which to compare the results of human modification. Innumerable laws of nature can never be thoroughly understood without some access to the conditions of the primeval. These scientific values are quite generally recognized, and both the Forest Service and the Park Service are making provisions for preserving them.

The importance of the primeval in the more subtle aspects of forest recreation is much less generally recognized. Yet observation of many recreationists gives convincing evidence that most of those who visit the forests for contemplation, inspiration, communion with nature, or enjoyment of the beautiful receive very much more pleasure from the beauty of primeval areas than from the beauty of areas modified by man. To argue that because lovers of the woods get some enjoyment from cut-over lands it is unnecessary to finance the retention of samples of the primeval would be as unconvincing as to argue that because lovers of music get some enjoyment from the concerts of mediocre pianists it is unnecessary to finance a Paderewski or a Rachmaninoff.

It seems necessary not only to reserve samples of the primeval, but to reserve a considerable number of such samples in every important forest type in the country. There are at least three important reasons for this: first, in order to avoid such overuse as would speedily destroy primeval conditions; second, to make the advantages of primeval areas readily accessible to people in as many different parts of each forest region as possible; and third, because there is still so much danger that any forest stand will be wiped out by fire or insects. Therefore, unless it is already too late, at least half a dozen primeval areas should be set aside in every forest type.

It is not reasonable to contend that if all the samples of primeval conditions in one sort of forest were destroyed other types would do as well. Each timber type has its own unique charms, its own unduplicated manifestation of beauty the destruction of which would distinctly lessen the potential esthetic enjoyment of nature. There are as great differences between the beauty of the redwood, the Engelmann spruce, the western white pine, the northeastern hardwood, the southern pine, and the cypress forests as there are between the paintings of Da Vinci, Turner, Rembrandt, Sargent, Goya, and Diego de Rivera. If anyone proposed that public expenditures for the housing of art should be reduced by disposing of the works of Turner, Sargent, and Rivera, because their paintings are less distinguished than those of the others, the suggestion would be greeted with uproarious derision, because everyone knows that art can not be relatively evaluated. Similarly no defensible rating can be made of the relative beauty of timber stands, and consequently each important type should be preserved in the museum of the forest.

The minimum area necessary for the maintenance of primeval conditions varies with forest type, climate, and topography. In general the Forest Service believes that for scientific purposes 1,000 acres is about the smallest area desirable, though in special cases where so much as

1,000 acres of virgin forest can not be found the largest available area will have to be sufficient. A tract of 1,000 acres, while well adapted for research, is too small for satisfactory recreation. The person with a yearning for the beauties of the primeval wants to do more than just stroll into a virgin stand of timber and squat. He desires to be able to walk around in it for a considerable period, losing himself for a while in its timeless beauty, forgetting that there is such a thing as a machine-age world. This is extremely difficult to do in 1,000 or even 5,000 acres. While the figure varies considerably with timber type, a primeval area should contain not less than 5,000 acres, which is, after all, scarcely 8 square miles. Thus far little progress has been made in reserving primeval areas of even the minimum size, and only the spruce forests of the Northeast and the lodgepole pine and subalpine forests of the West have had half a dozen of such tracts set aside. It is vitally important that any additional areas which are to be preserved should be set aside at once, for without early action the possibility of maintaining primeval areas in a number of important timber types will be lost.

"Wilderness areas" are regions which contain

no permanent inhabitants, possess no means of mechanical conveyance, and are sufficiently spacious for a person to spend at least a week of active travel in them without crossing his own tracks. The dominant attributes of such areas are: first, that visitors to them have to depend exclusively on their own efforts for survival; and second, that they preserve as nearly as possible the essential features of the primitive environment. This means that all roads, settlements, and power transportation are barred. But trails and temporary shelters, features such as were common long before the advent of the white race, are entirely permissible.

It will not be possible to preserve primeval forest conditions through the whole of any tract large enough to fulfill these requirements. Indeed there may be some wilderness areas, as for instance in Maine, where practically the entire tract will have been logged. The difference between primeval and wilderness areas is that the primeval area exhibits primitive conditions of growth whereas the wilderness area exhibits primitive methods of transportation. Of course wilderness areas may contain within their boundaries much that is primeval. Their chief function, however, is not to make possible contact with the virgin forest but rather to make it possible to retire completely from the modes of transportation and the living conditions of the twentieth century.

As regards area, Americans who want wildernesses of the sort which existed in frontier days will have to retreat to the more remote expanses of northern Canada and Alaska. Some semblance of pioneer conditions might be obtained in tracts of more than a million acres, of which several still remain uninvaded by roads. For those wishing to spend one or two weeks in wilderness travel without retracing their routes, 200,000 acres (about 300 square miles) is the smallest area that would generally prove satisfactory. In many cases smaller units have a great recreational value, but it is not the value of wilderness travel.

There remain today in the United States only 9 areas of 1,000,000 acres or more, 18 areas of 500,000 acres or more, and 38 areas of 200,000 acres or more which could still be set aside as wilderness tracts without involving any serious sacrifice of potential raw material, any great risk to adequate fire protection, or any major changes in existing highway plans. These tracts total

about 27 million acres, of which at least 7 million
acres is nonforested land, leaving about 20 million acres in a forested condition. The setting
aside of most of this area would involve relatively
little economic sacrifice because it lies chiefly
in unfertile and inaccessible mountain country
where commercial values are low. The main
sacrifice would be in barring automobiles. In
view of the fact that 486 million acres of forest
land in the United States would still be subject
to highway development, it would certainly
seem conservative to bar mechanized transportation from the scant 20 million acres of forest land
which alone remain to remind us of the vast
wilderness tracts of our recent pioneer history.

The great majority of people who today visit
the forests for recreation do so by automobile.
While most of these visitors do not penetrate
into the forest, many of them are very much
concerned with the part which they can see from
the highway. If this were destroyed or seriously
damaged their enjoyment of touring would be
immeasurably impaired. Consequently, it follows that for these people it is of great importance to preserve from serious scenic damage
the timbered strips adjoining the more impor-

tant roads. These strips will be referred to as "roadside areas." Upon them scenic values should not only be safeguarded where they already exist, but also restored through planting, the elimination of billboards, the tearing down of ramshackle houses, and the cutting of brush and timber to open attractive vistas.

The desirable width for these roadside areas varies with the density of the forest, the topography, and the danger of windfall. Along roads that are used chiefly in summer, when forest visibility is much less than in winter, the strips can be narrower. In most cases the width of the strips on each side of the road should probably range from 150 to 500 feet, which would be sufficient to hide any impairment of the scenery behind them.

For the benefit of many automobilists and boat travelers who spend their nights in camp, it is important to provide "camp-site areas" with fireplaces, water, and cleared spaces for tents. These will not only make camping more comfortable and convenient for travelers who are not out to rough it, but also curtail the very serious damage to forests which results when hordes of tourists who throng many popular highways camp chaotically through the surrounding woods. In general it is desirable to locate camp-site areas some distance back from the roadway. This spares the camper the continual dust cloud raised by passing automobiles, and spares the tourist constant interruption in his enjoyment of the forest. To a certain extent the location of camp grounds will be determined by the occurrence of water, though wells and preferably piped water will make almost any area available for this use. In many populous parts of the country practically every possible camp ground should be developed. The laying out of camp-site areas calls for landscape architecture and recreational planning of high quality.

"Residence areas" provide space for private homes, hotels, resorts, group camps, sanatoria, and stores and services of one sort or another. They are today in most cases privately owned, though the national forests in 1931 rented summer home sites to some 493,235 people and their guests. About one-quarter acre is the minimum desirable area for a single forest residence, while one acre would probably not be an unreasonable average. A store or service station generally needs a little less acreage than a private home, while a hotel or a group camp requires considerably more.

Although for many people the automobile tour in itself supplies every want, increasing numbers of people desire more intimate contact with the woods. Their first trip may be confined to the highway, but soon a curiosity as to what lies beyond the roadside fringe is roused, and they return to find out. Once the joys of intimate contact with the forest are discovered, the idea of taking a vacation exclusively on the highway seems almost as preposterous as it would seem to have moving sidewalks established in picture galleries so that one might enjoy the paintings without stopping. These people will generally be interested only in an afternoon's walk or an all-day hike or at most in an overnight trip. They do not require the large expanse of a wilderness, and may not have ready access to any superlative or primeval areas. What they need is any tract of forest on which they can get away from the sounds of the highway and which has not been severely injured scenically. This type of area will be referred to as an "outing area."

Anybody who visits the popular outdoor centers near the large American cities on public holidays cannot fail to be impressed by the immense number of people in search of forest recreation. Almost every possible open spot along the highways will be filled with picnickers and campers cooking their meals, pitching tents, playing games, swinging from trees, and noisily giving vent to pent-up childlike impulses which are rigidly curbed under city regimentation. At night time it sometimes seems almost as if every tree and every bush sheltered a sleeping figure. Such heavy use does more than kill the possibility of enjoying one of the chief recreational values of the forest, its privacy and quiet. It threatens to destroy the forest itself.

This damage to the forest comes from several sources. Small trees are cut down for fuel and tent poles. Reproduction and underbrush which gets in the way is removed. Trees and the lesser vegetation are killed by abrasion. Gasoline and oil leaking from cars onto the forest floor seriously injure all forms of plant life. Finally, the constant tramping on the forest floor kills the moisture-absorbing, nitrogen-providing ground cover, and packs down the soil so firmly that proper aeration for the roots of the trees becomes impossible. Many much-frequented camp grounds originally laid out in beautiful groves have become virtually deserts, with the few remaining trees all sickly, with the undergrowth and reproduction virtually extinct, and with the surface of the camp ground merely bare soil which in wet weather becomes a slimy mud and in dry weather gives off a constant cloud of dust.

The remedy for such intolerable conditions is fourfold. It is first of all essential to set aside sufficient camping ground in the vicinity of large centers of population to assure that no camp-site will be overused. How many camp-sites will be needed for each community is a subject for individual study, involving a consideration of the population, its present and probable future recreational habits, the quantity of other types of recreational land accessible for use, and finally the susceptibility to damage of each type of forest available for camp-sites.

A second important remedy is to educate the public in the proper use of camp-sites. There is an immense difference between the damage caused by careful, considerate people and that caused by people who appreciate nothing except their own immediate pleasure. Lack of regard for other campers is in most cases not wilful but simply thoughtless. Much of the misuse and destruction of public camp-sites could be eliminated by educating people into a consciousness of the necessity for care with fire, prevention of needless damage to trees and other vegetation, cleanliness in garbage disposal, and reasonable consideration of the peace of neighboring campers.

There is need also for careful planning of camp-sites so that cars may be parked, tents pitched, campfires built, and meals eaten at definite places instead of all over the camp ground. In this way the quantity of vegetation destroyed by camping can be greatly reduced and much more efficient use can be made of the available space. Enough well-planned camp-sites are already in use to indicate clearly how much better the forest values are protected on such areas.

Finally, it is desirable to have definite regulation of all camp grounds. In the case of large, much-frequented camp-site areas this will necessitate a resident administrative officer; in the case of smaller, less popular areas, an occasional visit by an inspector. The administrator will see that necessary regulations are enforced, he will help to educate the camp-ground users as to the desirability of good camping manners, he will assist them where they require help, he will build fireplaces, garbage receptacles, driveways, he may in some cases even procure fuel for campers, and he will clean up any mess they may leave behind and repair any damage that can be repaired.

It is not alone man who heaps destruction on our recreational areas, for there are also many natural enemies which if uncontrolled will certainly ruin the charm of the forest. Areas devoted to recreation require just as serious protective measures as do commodity areas; in fact fires, insects, and fungi may be even more destructive to recreational than to commercial values. Dead trees often have a salvage value as commodities, but there is no way of salvaging the destroyed beauty of a ruined forest.

It will be impossible to make any precise calculations of the area necessary to meet the country's forest recreational needs until some survey, much more thorough than anything yet

attempted, has been made. However, the following estimate by types of recreational land may
have some validity as a rough approximation:

Superlative areas	3	million acres
Primeval areas	9½	" "
Wilderness areas	20	" "
Roadside areas	4	" "
Camp-site areas	1½	" "
Residence areas	6	" "
Outing areas	11	" "
	——	
Total	55	" "

It is well to point out that 55 million acres
is only 11 percent of the total 506 million acres
of commercial timber land in the United States.
This seems like a very conservative area to devote to the purpose of assuring tens of millions
of our citizens a fitting environment for the finest moments of their lives.

CHAPTER XII

A Biological Interlude

IN THE Selkirk Mountains of northern Idaho are many primordial forests of towering white pines and larches and Douglas firs which overshadow an understory of densely crowned cedars and hemlocks and Engelmann spruces. As a person walks among the disordered trees and gazes upward through the many tiers of the dark needled canopy to where it brushes against the sky, he has the impression of a perpetual twilight and of almost infinite depth. He sees the most amazing variety around him: variety of tree trunk, from the light gray, deeply grooved, clean-boled shafts of the white pines, to the reddish, flaky-barked trunks of the hemlocks, with thick-set branches almost to the ground; variety of needle patterns, from the long-leaved whorls of the white pine to the light green rosettes of the larches and the dark, bristly spikes of the spruces; variety of size, from the tiny, one-year-old seedlings, which can hardly be noticed unless one gets down on hands and knees, through the intermediate hemlocks and cedars, up to the most gigantic white pines and larches, which sometimes rise nearly 200 feet and exceed fifty inches in diameter. The wind blows freshly, mysteriously, guardedly, through these great forests, and with it comes a feeling both of peace and of infinite complexity.

This complexity becomes especially vivid as one delves into the autobiographies which these forests have left off of their own development. Back in 1650, about the time that Cromwell was slaughtering the Scotch at Dunbar, a great fire, raging unimpededly in the Selkirk Mountains, wiped out all of the old stand on Squaw Creek except a few, thick-barked larches.[35] Apparently there must have been a good white pine seed year in 1650, because the fire was followed almost immediately by an abundant reproduction of white pine. In addition to this good seed crop at the time of the fire, we know definitely from the width of tree rings in neighboring forests that the years immediately following 1650 were wet ones in northern Idaho. Consequently, the new trees had very favorable conditions for survival and growth, and they speedily developed into a dense, young stand of white pine. This stand at first must have been so thickly stocked that very few less rapidly growing trees had a chance to get started. But after about fifty years, in common with most white pine forests, this one began to thin out, and the seeds of the shade-enduring cedar and hemlock must have been carried by wind and animals from neighboring clumps of timber which had not been affected by the fire. For the next 225 years, in quantities which fluctuated with the climatic cycle and the prolificness of seed source, these species continued to germinate under the main canopy of the forest and to grow in spite of the scant amount of sunlight that penetrated to them. In occasional openings, caused by the

death of large trees from lightning or insects or
windfall, additional white pines and larches and
a few Douglas firs seeded into the forest, but in
the dense shade under most of the stand these
species could not grow.

All through the centuries that this forest was
developing, a vivid epic of biological interrela-
tionship was being enacted. It was the interrela-
tionship of tree with tree, of trees with the lesser
vegetation, of trees with the slowly changing
soil, of trees with the continually varying climate, and of trees with insects, fungi, and their
own advancing senility. Portions of the story can
still be plainly read in the growth rings of the
trees, portions can be deduced from the forms of
the trees and the pruning of their branches, portions are obvious in the decaying trunks on the
floor of the forest, portions are told by the varying herbaceous vegetation, and portions can be
pieced together from a comparison of the soil
in this forest with that in similar stands of different ages. No drama of human life could be more
varied than this drama of the infinitely complex,
constantly changing forest.

In 1925 this stand was marked for cutting by
the Forest Service and logged by the Dalkena

Lumber Company. The Forest Service used the best silvicultural technique of which it had knowledge. It left several white pine seed trees on every acre, and besides this a goodly stand of the smaller hemlocks and cedars. However, there had been a poor white pine seed year in 1925, and 1926 was exceptionally dry, so the few white pine seedlings which germinated were killed by drought. As a result, no new trees of the valuable white pine species survived the first year after cutting, and I heard several people, judging the quality of the Forest Service's silviculture from that year alone, evaluate it with satirical scorn. However, in 1926 the white pines all bore a most prolific crop of cones, and the two following summers were unusually moist. Consequently, a great many white pine seedlings germinated and lived through their first two critical years. As a result, even though 1929 and 1931 were very dry, a sufficiently large number of white pine trees are now certain of survival to assure an excellent mature stand.

Meanwhile, not more than three or four miles away, another stand which appeared almost exactly the same and was handled by similar silvicultural methods, showed in 1931 almost no white pine reproduction. The reason may have been that the stand was growing on a gentle south slope and received enough additional sunlight to desiccate the soil just sufficiently more to prevent the white pine seedlings on this site from surviving the drouths of 1929 and 1931. But again, among the great complexity of factors involved, it is highly probable that the cause of the failure may have been some influence which as yet has gone entirely unobserved. This case illustrates the profound effect which minute differences of environment may have on the successful practice of forestry. It indicates how tremendously important is the detailed knowledge of tree habits if we are to maintain the productivity of our forest lands.

So subtle is the delicacy of nature's balance that a minute alteration of conditions may change entirely the type of forestry which should be practiced. Climate is continually fluctuating, and a method of cutting which may work excellently when it is followed by moist years, may not be applicable at all to periods of drouth. Methods of cutting which work on north slopes may be complete failures on south-facing ones, as has already been indicated. Silvicultural practice which may be splendidly successful on a light soil may be dismally inadequate on a heavy one. I recall two superficially similar New England white pine forests. Each was cut by the so-called shelterwood system in which the ground was left sheltered by about one-third of the original forest. In one which grew on light, sandy soil this worked splendidly, and a new forest of thrifty white pines is already well advanced under the open canopy of what remains of the old forest. This will soon be logged, leaving a dense stand of young pines which should yield practically the maximum growth possible on that site. On the other area, where the soil was heavy and well adapted to hardwood trees, the cutting of two-thirds of the original forest admitted enough light to permit a dense growth of white ash, red oak, white oak, gray birch, and soft maple to develop. These grew so much more rapidly than the white pines that they choked out the reproduction of this generally more valuable species. When the remainder of the original pine forest is cut on this heavy loam there will remain an almost pure hardwood stand.

These accounts of a few relatively obvious relationships suggest the immense intricacy of the forest. Even that basic unit of the forest, the tree itself, is so involved that 100 years of plant physiology have only taught us in the most general way how it functions. Shorn of all details and all potentially valid counter-hypotheses, the following would be a simple statement of the life processes of a tree.

As is true of all ordinary green plants, the tree's most fundamental activity is the combination of water from the soil with carbon dioxide from the air to form sugar, or some water-soluble substance closely resembling sugar, under the influence of sunlight. Oxygen gas is also produced, some of it being used in respiration while most of it escapes into the air. Some of the sugar formed by this photosynthetic process, which can occur only when light is available, is converted into cellulose and similar substances, and goes to make the walls of the minute cells that compose plant tissues. The cell walls divide the plant body into innumerable minute chambers or passages, of very many different shapes and sizes, but of just three basic sorts: living cells containing protoplasm, dead cells, and intercellular spaces. The two last mentioned chambers are regularly without living protoplasm, their cavities containing water with dissolved and suspended substances, or else just gas. All of the vital processes of a tree are carried on under the influence of the protoplasm of the living cells, which are continuously supplied with water, food, a number of essential minerals, and oxygen. Sugar formed in the green tissues not only gets altered to form cellulose and starch, but also gives rise to such compounds as fats or oils, and it gets combined with salts, derived from the soil, to form proteins and many other exceedingly complex organic substances. Sugars, fats, proteins, and related compounds, which are properly the cell foods, are used in the formation of additional protoplasm and new cell walls, in the production of resins, waxes, pigments, stored starch, essential oils, and numerous other characteristic substances. They are also used in respiration, an oxidation process, as in animals and man, which furnishes a source of energy for cell activity in general.

Sugar and other water-soluble foods are transferred from the places of their formation, in the green leaves, to all other parts of the tree. At the same time, water and dissolved mineral salts (including nitrogen, phosphorus, sulphur, potassium, calcium, magnesium, iron, oxygen and an uncertain number of additional chemical elements) are transferred from the absorbing surface of the roots. One of the most discussed of all natural phenomena is the ascent of water in trees. The plain facts concerning this are astounding in many instances. For example, water taken from fairly dry soil may be transported upward, sometimes to a height of more than three hundred feet, to the highest leaves, from which it is given off as water vapor into the surrounding air. In a single tree this may occur at the rate of many barrels per day. Not so much discussed, but surely just as interesting, is the corresponding fact that all the wood of a tree's root system must have been derived from sugar or other foods that have been brought downward from the leaves in spite of a much more rapid movement of water in the opposite direction.

This is merely an outline of what happens. It does not tell anything about how it happens. I will explain superficially what occurs to the best of our knowledge in just one vital process, the intake of water by the tree roots from the soil. Water is not absorbed by the roots which we

see when we dig in the ground around any tree,
but by very minute rootlets, ordinarily invisible
to the naked eye, which are called root hairs.
These root hairs are very thin-walled cells composed of three layers. The outside layer consists
of colloidal cellulose known as the cell wall. Inside of this is a sack of highly complex protoplasm. Inside of the protoplasm is the cell sap
consisting of water with dissolved minerals. The
water in the ground is absorbed by the colloidal
wall and comes into contact with the protoplasm.
This acts in the manner of a semi-permeable
membrane, in other words a membrane through
which some substances but not all can pass. It
is the characteristic of organic matter that when
two similar solutions are separated by a membrane which is permeable to the solvent but
not to the solute, the solvent from the more
dilute will flow through the membrane into the
more concentrated solution until the two are of
equal strength. If there were nothing limiting
this condition, the water from the dilute soil solution would flow through the protoplasm into
the concentrated cell sap until the solution became as dilute as that in the soil. This would involve an immense intake of water into the cell

and would mean a tremendous expansion in its size. Actually this process is limited by the pressure which the cell wall exerts, so that water flows into the cell only until the pressure of the expanding cell sap is just counteracted by the pressure of the cell wall. When this state of equilibrium is reached, the flow of liquid through the membrane ceases and the cell becomes saturated. Actually in nature this condition only obtains in aquatic plants. In land plants the water is drawn away from the root cells by an entirely different method from which it enters them, and is conducted up the stem of the tree with such rapidity that the intake of water never can exceed the outgo sufficiently for the cells to become completely saturated. This is a condensed physiological discussion of the intake of one substance which the tree needs, without any reference to the basic biochemical causes.

The known biochemistry of protoplasms alone fills hundreds of volumes, and what we know about it is negligible in comparison with what remains to be learned. Merely to give the various theories of why it acts as a semi-permeable membrane would more than equal in length all the remainder of this book. To discuss the structure of even a single one of the many proteins contained in protoplasm would be a staggering task, without considering the numerous protein decomposition products, such as the amino acids or the polypeptides, and without considering such other protoplasmic substances as the alkaloids, the carbohydrates, the glucosides, the pectins, the tannins, the pigment substances, the fats, the lipides, the phosphatides, the vitamins, and the enzymes.

But protoplasm is only one of many biochemical problems, and water intake is only one of innumerable physiological processes which are going on in the tree. The tree itself is only one out of millions of similar units which must live in competition or cooperation with each other, with the lesser vegetation, and with an infinite variety of animal life from the submicrosopic bacteria in the soil to the deer, elk, and grizzly bears. All these factors are further complicated by the constantly changing climate and soil conditions which make the established relationships of one moment absolutely inapplicable the next. In order to understand all these interacting forces and thus be able to handle intelligently the problems of the forest, a vast amount of knowledge is necessary. Only through the slowest and most painstaking research, through hacking off little corners of the problem and working on them patiently and courageously, year after year, can the forester ever hope to acquire an adequate understanding of what he is working with.

The progress has been notable considering the overpowering disadvantages under which forest research has had to develop. The problem has been staggeringly complex, funds have been woefully insufficient, and the very necessity of pursuing most of the work in an outdoor laboratory has involved difficulties unknown to most research. Furthermore, forest scientists have generally been looked down upon as hopelessly lowbrow by the workers in the more developed and exact fields of science, while at the same time they have been considered hopelessly highbrow by most of the practical woodsmen.

Nevertheless, we now know the minimum requirements for the protection of all of the ma·jor forest types from actual devastation. We are able to predict how great a volume of wood a fully stocked, even-aged forest can produce at different ages and on different sites, though it is true we have a lot to learn about applying this information to the irregular stands found in nature. We know a little about the effect of different silvicultural operations, such as thinnings, weedings, release cuttings, and prunings. We have worked out certain very important correlations between climatic factors and forest inflammability. We are beginning to learn a great deal about the nature of the disastrous dry lightning storms of the West. We have made immense advances in the development of softwood nursery practice and planting technique. We know enough about the habits of a few insects and fungi to keep them in satisfactory control. Furthermore, through research we have revolutionized the production of turpentine from the forests.

Without going into details, the following are a few of the most important problems we must yet solve:

We must learn a great deal of detailed physiology and ecology about every important species of trees in the country so that we may know how to cut our stands in a way that will permit reproduction of the forest—avoiding not only devastation but also deterioration in the quality of the new forest. This knowledge must be complete enough for us to attain this goal not just in a majority of cases but with consummate regularity.

We must understand how to carry these reproduced stands to maturity: at what ages they need to have fast-growing inferior species weeded out, under what conditions they need to be thinned, if and how pruning operations should be carried out.

The inherited quality of trees varies as widely as does that of the cultivated plants. By the study of forest genetics we should eventually be able to grow in our nurseries a far better quality of trees than we produce at present. In this connection, the possibility of producing hybrids more valuable and faster-growing than existing species has tremendous potentialities.

We must continue to improve our nursery practice through studying ways of lessening the shock in transplanting and thus enabling trees to recover more rapidly after they have been planted. Especially we should devote attention to hardwood nursery practice, which is far behind that of softwoods.

Yield tables, which are estimates of how much

wood can be raised in specific types of forest, are at present very unsatisfactory for uneven-aged stands. In addition we have barely touched the surface in applying the estimated yields of timber which our theoretical tables indicate to the infinitely varying conditions which actually obtain in the forests.

While we have made splendid progress, there is an immense amount of additional information we need in order to be able to make effective use of weather predictions in battling forest fires. Studies of the relative inflammability of different natural fuels found in the forest are needed in order that we may find ways of eliminating the most hazardous ones.

In order to control the disastrous floods and to reduce the appalling rate at which soil is being washed away we require huge scale investigations all over the country on the relationship of forest and range to streamflow and erosion.

Very little has been definitely ascertained about the influence of the forests on climate. One observes the unusually dry weather over most of the United States during recent years and wonders if it could be related to the destruction of our forests. If there is any relationship, it is something which we should surely know. To obtain this knowledge will require very careful and original research.

We must study the detailed life histories of hundreds of different insects and fungi which are causing major damage to the forests. Only in this way can we hope to develop methods of controlling their ravages, both through understanding at which stages in their life history they are most susceptible to injury and through finding the easiest ways of changing their environment in a manner which may destroy them. With such knowledge we can devise silvicultural methods which will be as unfavorable as possible for insect development.

From what has been said in this chapter, and from a very great deal more that might have been said, it is clear that the practice of forestry requires much more than merely setting aright the economic and social machinery. Adequately managed public ownership and a high grade of intelligent social planning are essential background conditions, like a canvas on which a picture is to be painted, but the picture still has to be painted after the canvas has been suitably prepared. For this to be consummated the artist also requires paints and brushes. In our forest picture these take the form of specific scientific knowledge of the nature and behavior of trees and their surroundings and a delicate appreciation of the infinitely complex interrelationships of the forest which can not be reduced to science. With the background of a well planned and smoothly operating social and economic system, and with the paints and brushes of scientific knowledge and intuitive appreciation, the skilled forester, exercising a technique not unlike that prevailing in the better known forms of art, will create the beauty of the managed forest.

CHAPTER XIII

A National Program

UNDER their present management the American forests are drifting into constantly expanding ruin. Year by year the area of devastated land keeps mounting, until today it has reached the appalling total of 83 million acres, to which nearly a million acres are being added annually. Even more serious than the devastation is the grave deterioration which has occurred on at least 200 million additional acres. Between devastation and deterioration the

American forests and all the social values which they represent are indeed in a tragic condition.

The major cause of this sorry plight is the mismanagement of privately owned forests. Fire damage, erosion, devastation, and destruction of scenic values are many times more severe on private than on public forests. This hopeless insufficiency of private ownership has obtained even though the government has carried five sixths of the burden of fire protection. When this miserable failure is contrasted with the splendid record of public forest management the moral seems inescapable. Public ownership is the only basis on which we can hope to protect the incalculable values of the forests for wood resources, for soil and water conservation, and for recreation. It is urged, therefore, that the public should acquire at least 562 million acres out of the 670 million acres of potential forest land.

Regardless of whether it might be desirable, it is impossible under our existing form of government to confiscate the private forests into public ownership. We can not afford to delay their nationalization until the form of government changes, because if we do the forests will be so deteriorated as to be scarcely worth owning. Consequently, it will be necessary to acquire them by purchase. This purchase program cannot be consummated all at once. Pending the change from private to public ownership it will be necessary to regulate the private use of forests so that they are not too severely gutted by the time the government takes over their management. In spite of this genuine value of public regulation, it will only be an incidental to the main solution of our forest problem. This main solution, as has been stated, is public ownership. But public ownership is not a panacea. It must be backed by careful land planning, protection of the rights of those who labor on the forests, reorganization of rural government and redistribution of rural population, safeguarding of recreational values from commercial exploitation, and a great increase in the knowledge of the forest through an ambitious program of research.

The administration of the vastly increased federal domain might well be patterned after the splendidly efficient administration of the present Forest Service. It would, of course, be necessary to create many new national forests, and there would be a great increase in the work connected with land acquisition. These, however, would merely be changes in detail which would not alter the basic form of organization.

The activities of this expanded Forest Service would fall into four major fields. These would be:

Management, including the administration of protection, silviculture, recreation, grazing, and engineering projects;

Land acquisition and land planning;

Regulation of private forests;

Research.

It is well to consider the cost of the proposed forest program under each of these activities. The land acquisition is estimated to involve payment of about 41 million dollars annually for 25 years, figuring four percent of a total investment of $1,017,000,000. This will probably be enough to purchase the 240 million acres which ought to be transferred from private to public ownership. The four percent annual payments should be guaranteed for a five-year period as a special inducement to private owners to sell immediately. Thereafter, only that proportion of the payment should be met which can be raised from the net earnings of the public forests. In addition to this purchase price, an annual outlay of about one million dollars would be needed for several years to handle the work of appraising the lands to be purchased and clearing titles.

The Forest Service has estimated an average annual cost of 12 cents an acre as necessary for the adequate management of the National Forests.[36] Since these are in general the most inaccessible and difficult to administer of any of the forest areas in the country, it would not be unreasonable to apply these same figures to the private lands which would be more accessible but which would acquire a greater amount of reconstruction work. With 562 million acres of proposed public forests, this would involve a maximum annual outlay of approximately 67 million dollars. This sum would not be required until the full purchase program was realized. Thereafter it would decrease steadily as more and more areas were brought under forestry practice, for it is far more expensive to build up a deteriorated area than to continue the management of a tract in good condition.

The cost of regulation, once the acquisition program was realized, would be relatively small because the only lands which would have to be regulated would be the 80 million acres of farm woodlands and the 28 million acres of brushland which will probably remain in private ownership. The cost for this administration would not amount to more than 5 cents an acre or perhaps half a million dollars. At first, pending the consummation of the acquisition program, the cost of regulation would be considerably higher, but this would be more than balanced by the lower cost of administering the smaller area of public forests.

The necessary government expenditure for an adequate research program would be about 8½ million dollars annually. This would include investigations in silviculture, protection, grazing, forest products, and economics.[37]

The annual expenditure for management, regulation, and research would thus amount to approximately 75 million dollars. It will only be for a few years that this will constitute an expenditure, however, for shortly the profits of forest management should far exceed the costs. Furthermore, so many extraneous values will be derived from this outlay that even if it were not absolutely necessary to save our forests, and even if it were not self-liquidating, it would be largely justified.

In the first place, the greater part of the 75 million dollars would go directly into wages. At the present time, when there is such a tremendous need to increase the purchasing power of the country through putting as many people as possible to work, this expenditure would be a material help toward improving our economic condition. In line with this reasoning, the government is planning to pump $3,300,000,000 into circulation through public works. It would certainly seem that the adequate management of our forests is a public work worth a fortieth part of this vast sum.

At present it is generally recognized that the government must provide relief for a large share of the 12 million unemployed. It seems to be a serious problem to find enough useful work for these people to do. If the government were to take over the majority of forest lands there would be enough work to keep several million people busy for many years. The appropriations necessary to pay for this work would in reality not be costs chargeable against the forests at all, because the government would have to pay relief money to these workers regardless of whether they sat outside of the urban relief headquarters or whether they spent their time building up the forest wealth of the nation.

It is also important to consider that every year which we delay spending money to rehabilitate our forests will mean a greatly increased ultimate cost. The deterioration of the forest is proceeding at a geometric ratio, so the sooner we put an end to it, the less we shall have to pay. Since the cost will be higher later, the only sensible way is to protect sooner, regardless of the amount which past mismanagement now demands that we must spend.

However, it must be stressed again that the expenditures we make for forestry do not involve an ultimate drain on the public treasury. Once the purchase program and the early steps of rehabilitation are consummated, the forests will yield a far higher revenue than the expense of administering them. Kotok has made estimates on the potential costs and returns of the American forests. He says: [36] "From the foregoing conservative estimates of returns from forests handled under methods which will insure high rates of productivity, it may be concluded that the forests of the United States, maintained at a level of productivity sufficient to meet fully the national requirements, both for timber and other services, should produce a gross return of about $700,000,000. This is in terms of stumpage values, with a partial allowance for recreational, watershed, and other more or less intangible values. After the costs are deducted from the gross returns, there remains a net of $400,000,000 to $500,000,000 as an earning on the investment in all forest properties, public and private."

In order to carry out the program of public ownership and administration which has been recommended, certain specific legislation will be needed. The following is of major importance:

1. Congress and the state legislatures should first grant authority and then appropriate funds for the purchase of the 240 million acres which have been recommended.

2. Congress and the state legislatures should make the necessary appropriations for the adequate administration of the 562 million acres of proposed public forests.

3. Congress should add to the National Forest

system the 22 million acres of public domain forest lands. The government's mismanagement of these unadministered acres is the only blot on its otherwise excellent record of forest protection.

4. Laws should be passed by the federal government and the states which would automatically make tax-delinquent lands a part of the state forests if the states desire to administer them and otherwise a part of the federal forests.

5. Congress should increase the appropriations for federal research by about 1½ million dollars.

6. For those forest lands which will remain in private ownership, the federal government should pass laws giving itself the right to control fire and stop logging and grazing practice which leads to devastation.

There are two possibilities which stare us in the face with exceptional clarity. We can continue our present policy of being more solicitous of the rights of private timber owners than of the welfare of the public. There can only be one result of such a policy, the hopeless deterioration of our vitally needed forests. On the other hand, we can spend the large sums of money necessary to acquire and administer the public forests for the benefit of all the citizens. This policy will, in the long run, save us vast amounts of money, and it will preserve the unassessable value of the forest for timber production, for water and soil conservation, and for recreation. The time has come when we must discard the unsocial view that our woods are the lumbermen's and substitute the broader ideal that every acre of woodland in the country is rightly a part of the people's forests.

References

1. Marsh, R. E., and Gibbons, W. H. *Present Potential Timber Resources*. A National Plan for American Forestry. 1933.
2. Hallauer, Frank J. *Our National Timber Requirements*. A National Plan for American Forestry. 1933.
3. Chapline, W. R., and Campbell, R. S. *Forest Ranges*. A National Plan for American Forestry. 1933.
4. Roberts, Paul H., and Stone, J. H. *Wild Life a Forest Resource*. A National Plan for American Forestry. 1933.

[221]

5. Dayton, W. A. *Minor By-products of the Forest.* A National Plan for American Forestry. 1933.

6. Zon, Raphael, and Sparhawk, W. N. *Trends in Wood Consumption.* A National Plan for American Forestry. 1933.

7. Munns, E. N., and associates. *Watershed and Related Forest Influences.* A National Plan for American Forestry. 1933.

8. Bennett, H. H. *Unpublished Memorandum.* 1933.

9. Bennett, H. H. *The Quantitative Study of Erosion Technique and Some Preliminary Results.* The Geographical Review. July, 1933.

10. Marshall, Robert. *The Forest for Recreation.* A National Plan for American Forestry. 1933.

11. Anonymous. *Wild Life Conservation.* Senate Report No. 1329, 71st Congress. 1931.

12. Merriam, John C. *Educational Values of Recreation.* Educational Record. October, 1932.

13. Marshall, Robert. *The Problem of the Wilderness.* Scientific Monthly. February, 1930.

14. Clapp, Earle H. *Major Problems and the Next Big Step in American Forestry.* A National Plan for American Forestry. 1933.

15. Kirkland, Burt P. *Status and Opportunities of Private Forestry.* A National Plan for American Forestry. 1933.

16. Olmsted, Frederick E. *The Work Ahead.* Journal of Forestry. March, 1919.

17. Preston, John F. *Silvicultural Practice in the United States During the Past Quarter Century.* Journal of Forestry. March, 1925.

18. Munger, Thornton T. *Timber Growing and Log-*

ging Practice in the Douglas Fir Region. U. S. Department of Agriculture Bulletin 1493. 1927.

19. Anonymous. *Accidents in Manufacturing Industries, 1926–1930*. Monthly Labor Review, Bureau of Labor Statistics, May, 1932.

20. Sparhawk, W. N. *Public Regulation of Private Forests*. A National Plan for American Forestry. 1933.

21. Mosher, William E. and Finla G. Crawford. *Public Utility Regulation*. Harper & Brothers. New York. 1933.

22. Russell, Charles A. *The Public Utility Crisis in This Election*. National Popular Government League, Bulletin 159. 1932.

23. Rauschenbush, Stephen. *The Power Fight*. New · Republic. New York. 1932.

24. Mosher, William E., and associates. *Electrical Utilities*. Harper & Brothers. New York. 1929.

25. Ahern, George P. *Forest Bankruptcy in America*. Green Lamp Library. Washington. 1932.

26. Behre, C. Edward, and Munns, E. N. *The Area Which Can and Should Be Used for Forestry*. A National Plan for American Forestry. 1933.

27. Marsh, R. E., and Gibbons, W. H. *Forest Land the Basic Resource*. A National Plan for American Forestry. 1933.

28. Compton, Wilson. *Lumber Code Hearings at Washington, D. C.* July, 1933.

29. Cole, G. D. H. *The Essentials of Socialisation*. The New Fabian Research Bureau. London. 1931.

30. Kneipp, L. F. *The National Forests as a Form of*

Federal Aid to the States. A National Plan for American Forestry. 1933.

31. Hendrickson, C. I. *The Agricultural Land Available for Forestry.* A National Plan for American Forestry. 1933.

32. Wisconsin Statutes, Section 59.97.

33. United Railway and Electric Company of Baltimore v. West et al. 280 U. S. 234. 1929.

34. Roosevelt, Franklin D. *Looking Forward.* John Day. New York. 1933.

35. Marshall, Robert. *The Life History of Some Western White Pine Stands on the Kaniksu National Forest.* Northwest Science. June, 1928.

36. Kotok, E. I., and associates. *Ownership Responsibilities, Costs, and Returns.* A National Plan for American Forestry. 1933.

37. Clapp, Earle H. *Research in the United States Forest Service, A Study in Objectives.* A National Plan for American Forestry. 1933.

NOTE—*A National Plan for American Forestry,* from which nearly half of these references are taken, is Senate Document No. 12, 73d Congress, 1st Session.

Index

Alabama, black gum forests of, 26

American Automobile Association, on motor camping and travel expenditures, 66

Bankruptcy, in local government, 165–68

Barrels, 29

Beech, a cooperage wood, 30

Beer Act, and cooperage, 29

Bennett, H. H., on erosion of farm lands, 46–47, 48

Birch, a cooperage wood, 30

Black gum, growth of, 26

Bolts, production of, 25, 31

Brush, a fire trap, 95–96

By-products, 27, 31–32, 34, 39

California, lumber production of, 19

Camp-site areas, 181–82, 185, 186, 188

Canada, forests of, 40

Capper, Arthur, his bill for government regulation, 110

Cattle, on forest ranges, 33, 34

Cedars, for shingles, 31; logging of, 95; in Selkirk Mountains, 189, 191, 193

Cellophane, 38

Cellulose, 39–40

Chapline, W. R., on forest grazing, 34

China, 77

Civil service, 164–65

Clarke-McNary Act (1924), 90, 110

Climate, and silviculture, 194; influence of forests on, 205

Collective bargaining, 164

Colvin, Verplanck, 17

Community, and forest management, 80, 83–84, 119; effect of

225

forest nationalization on, 153–54
Composition roofings, 31
Confiscation, of forest land, 141–42
Conservation, of forests, 20; soil and water, 24, 43–55; movement, 88
Consumers, and forest management, 80, 82, 103, 119
Cooperage, production of, 25, 29–30, 32, 37, 38
Crawford, Finla G., 113
Creosote, as a preservative, 28, 29
"Cut out and get out," in lumber industry, 101
Cutover land, fires on, 95, 96; reproduction on, 97

Death rate, in lumber industry, 100
Department of Agriculture, U. S., 46
Department of Labor, report on industrial death rates, 100, 101
Depression, and consumption of wood, 35, 36, 104, 105
Destruction, at popular camp sites, 62–63

Diameter, and merchantability, 25, 26
Disease, forest destruction by, 33
Douglas fir, logging of, 32, 96–97; in Selkirk Mountains, 189, 192
Drought, forest destruction by, 33
Duff, destruction of, 50; nature of, 51–52; preservation of, 97
Durability, 29, 30, 31

Employment, in forest work, 81
Engelmann spruce, 189
Erosion, prehistoric, 16; following forest destruction, 20; problem of, 43–55; control of, 112, 143, 147
Escape, the forest as a place of, 72
Europe, lumber supply of, 40; public regulation of forests in, 111–13
Expenditures, motor camping and travel, 66–67
Exploitation, commercial, of recreation areas, 62

Famine, danger of a timber, 124

Farmlands, erosion of, 46–49; abandoned, 127, 142, 148, 151, 162

Fence posts, production of 24, 28–29, 32, 37, 38

Fiber board, 38

Fires, forest, 16; protection from, 17; in southern pineries, 18; drain of, on forest, 33; and erosion, 50, 51; their threat to recreation areas, 62; subsidy against, 90; in public *vs.* private forests, 92–98; public regulation and, 121; public ownership and, 125–26

Fishing, in forest, 34

Floods, 1927 Mississippi River, 44; lessened by forests, 53; control of, 112

Floors, without wood, 36

Forage, in forests, 24, 33–34

Forest, primeval, 13–16; devastation of, 16–22; second growth, 19; raw material from, 23–42; value of, for soil and water conservation, 24, 46, 50–55; value of, for recreation, 24, 57–76; management of, 77–85, 133–39; government acquisition of, 141–57; a national program for the, 209–219

Forest Service, record of, 20; supports subsidy to private owners, 92; on forest devastation, 92–94; logging practice of, 96, 97; regulations of, 120; provisions for primeval areas, 174; silviculture by, 192–94; expansion of, 212–13

Fuelwood, production of, 24, 27–28

Fungi, protection from, 17; and railroad ties, 28

Furniture, 30, 31, 38

Girdling, 32, 96

Grazing, in forest, 33–34, 50, 51; as cause of erosion, 49–50; public regulation of, 121

Greeley, William B., 91

Gum, destruction of trees for, 16

Hardwoods, for cooperage, 30; growth of, 195

Health, value of forests for, 68–69

Hemlock, as pulpwood, 29, 32; logging of, 32, 95; in Selkirk Mountains, 189, 190, 191, 193

Highways, through forests, 180–81

Holly Springs, Miss., study of erosion, 45

Hough, Franklin D., 17

Houses, prohibition of wooden, 36

Hunting, in forest, 34

Idaho, white pine stands of, 26, 95, 189–94

Indians, their attitude toward land, 15, 16

Insects, protection from, 17; forest destruction by, 33

Insull, Samuel, 115

Interior finishing, without wood, 36

Interstate Commerce Commission, 116

James, William, 70

Jefferson, Thomas, influence of forest on, 70

Kansas, case of erosion in, 48

Kneipp, L. F., on local benefits of forest nationalization, 154

Knots, 26

Kotok, E. I., on costs and returns of forests, 216–17

Labor, decrease in hours of, 63–64

Laboratory, forest as a, 70, 202

Land, use of, and forest management, 80–81, 159–62

Larch, logging of, 95; in Selkirk Mountains, 189, 190, 192

Lodgepole pine, stands of, 26

Logging, and erosion, 50; of small trees, 99; public regulation of, 121; private practice in, 137

Logs, production of, 25, 31

Lumber, production of,

24–27, 31, 36–38, 40–42

Lumber code, hearings on a, 100, 120

Lumber industry, history of, 17–19; and taxation, 83–84; working conditions in, 100–102; regulation of, 110–21; and forest nationalization, 157–58

Lumberjack, working conditions of, 100–102

Lumbermen, and public welfare, 83–84, 87–107, 110, 118

Machine age, nervous strain of, 64, 72, 177

Machinery, in lumber industry, 19

Maine, as a lumber state, 17, 135, 178

Maple, a cooperage wood, 30

Maple sugar, 35

Masonite, 38

Merchantability, 25–26

Merriam, John C., on a national park vacation, 70

Michigan, as a lumber state, 18

Mills, public ownership of, 155–56, 158

Mine timbers, production of, 25, 30

Minnesota, as a lumber state, 18

Mississippi River, 1927 flood of, 44; silt in, 45

Monopoly, in lumber industry, 130

Mosher, William E., 113, 134

Mosher and Crawford, on public regulation, 113–14

Muir, John, 70

National forests, visitors to, 58, 60–62; management of, 132–33

National monuments, visitors to, 58

National parks, visitors to, 58, 60–62; vacationing in the, 70; as "superlative areas," 172–73

National Recovery Administration, 120

Nationalization of forests, 141–58

Naval stores, 34–35

New England, logging white pine in, 195

New York, as a lumber state, 18, 135

Northeast, spruce-hardwood forests of, 30, 32

Olmsted, Frederick E., on cooperation with lumbermen, 89

Ontario Hydro-Electric Commission, 134

Oregon, lumber production of, 19; public utilities in, 115

Outing areas, 188

Overcapitalization, in lumber industry, 103

Overproduction, in forest industry, 129, 157

Pacific Northwest, logging in, 96–97

Park Service, provisions for primeval areas, 174

Parks, county and municipal, visitors to, 59; See also National parks, State parks and forests

Pasture lands, erosion of, 46, 49–50

Pennsylvania, as a lumber state, 18

Pinchot, Gifford, for public regulation of forests, 88, 109, 110

Pineries, southern, 18, 32

Pines, as pulpwood, 29

Pioneers, destruction wrought by, 16, 17; tradition of, 71–72

Planning, national, and forest policy, 99–100; rural, 132

Plywood, 30

Population, redistribution of rural, 166–67

Power commissions, 114, 115

Power industry, public vs. private ownership of, 134–35

Preservatives, 28, 29

Prices, lumber, 102, 103; forest land, 147–48

Primeval areas, 173–77, 188

Private ownership, of forests, 84–121, 132, 143, 144–48

Prohibition, end of, 38

Protection forests, 111–12

Public forests, early advocates of, 17; management of, 20, 105

Public ownership, of forests, 84–85, 106–107, 121–41, 143, 210

Public regulation, of forests, 84, 109–21, 130, 212, 213, 214, 218

Public utilities, regulation of, 113–16

Pulpwood, production of, 24, 32, 37, 38

Railroads, use of ties by, 24, 28, 37, 38; regulation of, 116

Rangers, forest, 133

Rauschenbush, Stephen, on power industry, 134

Raw material, from forests, 23–42

Rayon, 38

Recreation, use of forests for, 24, 34, 57–76, 112, 124–25, 171–88

Reinvestment, by lumbermen, 91, 103

Research, in silviculture, 202–207, 212, 214, 218

Residence areas, in forests, 182, 188

Resin, and pulpwood, 29

Road-side areas, 188

Rocky Mountains, lodgepole pine stands of, 26

Roosevelt, Theodore, 88, 109

Rosin, production of, 34–35

Russia, coniferous timber of, 40

Salesmanship, high pressure, 130

Schurz, Carl, 17, 109

Seed trees, few left by lumbermen, 18; Forest Service and, 96

Selkirk Mountains, forests of, 189–94

Senate Committee on Conservation o f Wild Life Resources, 66

Sheep, grazing in forest, 33–34

Shelterwood system, 195

Sherman Anti-Trust Act, 129, 130

Shingles, production of, 25, 31, 37, 38

Silt, in rivers, 45, 54

Silviculture, in Northeast, 30; by-products of, 31–32; by private owners, 91

Society of American Foresters, favors subsidy to lumbermen, 90

Softwood, famine in, 40

Soils, 195

South, pineries of, 18–19; tree growth in, 26

Southeast, region of fastest growing timber, 26; pine stands of, 32

Sparhawk, W. N., on European forest regulation, 111–12

Spruce, chief source of pulpwood, 29; forests, 30; logging of, 32

State legislatures, lumber industry and, 87, 110

State parks and forests, visitors to, 58

Streamflow, influence of forests on, 51–55, 143

Subsidy, to lumbermen, 84, 90, 98, 105, 134

Superlative areas, 172–73, 188

Supreme Court, ruling of, on profit, 114; attitude of, towards public regulation, 117

Switzerland, public regulation of forests in, 111, 113

Tanning extract, destruction of trees for, 16; production of, 35

Tax delinquency, 147, 149, 152, 218

Taxation, lumber industry and, 83–84, 90, 106, 129, 145, 147

Thinning out, 28, 39

Thoreau, Henry D., 70

Ties, production of, 24, 28, 37, 38

Timber, virgin, 19, 22, 74, 173–77

Towns, abandoned, 83

Transpiration, in trees, 53–54

Transportation, increased ease of, 64

Trees, physical nature of, 53; planting of, 127, 133; life processes of, 196–202

Turpentine, production of, 24, 34, 35

Twain, Mark, 70

Unemployment, relief of, 215

United States, wood resources of, 33; decrease in wood trade, 40; its advantages as a wood grower, 41; floods in, 44; erosion in, 45–47

Utopianism, 79

Veneer logs, production of, 25, 30, 31

Wages, in lumber industry, 100; after forest nationalization, 162–64

Washington, lumber production of, 19

Watershed protection, 94, 111, 143

West Coast, lumber industry on, 19, 32

White fir, logging of, 95

White pine, growth of, 26; forests in Idaho, 95, 189–94; in New England, 195

Wild life, and forests, 24

Wilderness areas, 177–80, 188

Wind, forest destruction by, 33

Wisconsin, lumber production of, 18; stud-

ies of erosion in, 45; public utilities in, 115

Wood, American consumption of, 35–42; world consumption of, 40–41; substitutes for, 104, 131

Woodlots, farm, 27, 121, 143, 144, 148

Workers, and forest management, 80, 81–82; condition of, in lumber industry, 100–102, 119, 132; government forest, 131; and forest nationalization, 158, 162–65

Zon and Sparhawk, on world situation in wood, 40